Flight dynamics and system identification for modern feedback control

Flight dynamics and system identification for modern feedback control

Avian-inspired robots

JARED A. GRAUER
AND
JAMES E. HUBBARD, JR.

WOODHEAD PUBLISHING

Oxford Cambridge Philadelphia New Delhi

Woodhead Publishing Limited, 80 High Street, Sawston, Cambridge, CB22 3HJ, UK
www.woodheadpublishing.com
www.woodheadpublishingonline.com

Woodhead Publishing, 1518 Walnut Street, Suite 1100, Philadelphia, PA 19102-3406, USA

Woodhead Publishing India Private Limited, G-2, Vardaan House, 7/28 Ansari Road, Daryaganj, New Delhi – 110002, India
www.woodheadpublishingindia.com

First published in 2013, Woodhead Publishing Limited

British Library Cataloguing-in-Publication Data

A catalogue record for this book is available from the British Library

Library of Congress Control Number: 2013931613

Woodhead Publishing ISBN: 978-0-85709-466-7 (print)

ISBN: 978-0-85709-467-4 (online)

Typeset by Domex e-Data Pvt. Ltd., India.

Printed in the UK and USA.

For my family and friends

"There is a theory which states that if ever anybody discovers exactly what the Universe is for and why it is here, it will instantly disappear and be replaced by something even more bizarre and inexplicable. There is another theory which states this has already happened."

-Douglas Adams

Contents

List of figures

List of tables

Nomenclature

Throughout this document, scalar mathematical symbols are represented by lower case characters, vectors by lower case bold characters, and matrices by upper case bold characters. Differentiation of a scalar quantity by a column vector results in a row vector.

Roman Symbols

$\mathbf{A}$, $\mathbf{B}$	linear system matrices
AR	aspect ratio
$\mathbf{a}(\mathbf{p}, \mathbf{v})$	generalized aerodynamic forces
a	acceleration vector
b	wing span
$\mathbf{C}(\mathbf{p}, \mathbf{v})$	dynamic coupling matrix
C_{ij}, K_{ij}	center of mass location and reference frame
C_x, C_y, C_z	force coefficient components
C_l, C_m, C_n	moment coefficient components
$\mathcal{C}$	controllability matrix
$\bar{c}$	mean aerodynamic chord
$\mathbf{e}_x$, $\mathbf{e}_y$, $\mathbf{e}_z$	elementary unit vectors
$\mathcal{F}\{.\}$	Fourier transform
f	frequency
f_f, $\mathcal{T}_f$	flapping frequency and period
$\mathbf{g}(\mathbf{p})$	generalized gravitational forces
g	acceleration vector due to gravity

$\mathbf{h}$	magnetic field vector
$\mathbf{I}$	inertia tensor
$\mathbb{I}$	identity matrix
$\Im\{.\}, \Re\{.\}$	imaginary and real parts
$\mathbf{J}_f$	force Jacobian matrix
$\mathbf{J}_\eta$	attitude Jacobian matrix
$J(\phi)$	cost function
j	imaginary number
L, M, N	body-fixed moment components
l, r	in-board and out-board position vectors
$\mathbf{M}(\mathbf{p})$	generalized mass matrix
m	scalar mass
n_p, n_v	number of position and velocity states
$\mathbf{p}$	generalized position vector
p, q, r	rotational body-fixed velocity components
Q	dynamic pressure
$\mathbf{R}$	rotation matrix
R^2	coefficient of determination
Re	Reynolds number
$\mathbf{r}$	position vector
S	reference area
$S(.)$	skew operator
$T(\mathbf{p}, \mathbf{v})$	kinetic energy
t	time
$U(\mathbf{p})$	potential energy
$\mathbf{u}$	perturbation control input vector
u, v, w	translational body-fixed velocity components
V	airspeed
$\mathbf{v}$	generalized velocity vector
$\mathbf{W}$	weighting matrix
X, Y, Z	body-fixed force components
$\mathbf{x}$	perturbation state vector
x, y, z	Cartesian position components
$\mathbf{y}$	model output vector
$\mathbf{Z}, z$	rotation axis matrix and vector
$\mathbf{z}$	physical state vector

Greek Symbols

α	angle of attack
β	sideslip angle
Γ	Hamel coefficient matrix
Δ	increment, perturbation
$\delta_w, \delta_{lon}, \delta_{lat}$	wing and tail control inputs
δ, ϵ	quaternion scalar and vector parts
η	orientation vector
θ	vector of articulated joint angles
λ, **v**	Eigenvalue and Eigenvector
μ	physical control input vector
ν	translational velocity vector
ρ	density
Σ, σ^2	covariance matrix and scalar
τ	generalized force
$\Phi(t, t_0)$	state transition matrix
Φ, Ψ	kinematic Jacobian matrices
ϕ	parameter vector
ϕ, θ, ψ	roll, pitch, and yaw Euler angles
ω	angular velocity vector
ω	radian frequency

Superscripts

$^{-1}$	matrix inverse
^{T}	matrix transpose
$\dot{}$	time derivative
$\check{}$	ensemble average
*	trim value
$\hat{}$	estimated value

Subscripts

$_{ij}$	linkage i on kinematic chain j
$_{m}$	measured value

Other

$\lVert . \rVert_p$	p-norm
$\otimes$	Kronecker product

Acronyms

CAD	computer aided design
DC	direct current
EE	equation-error
IMU	inertial measurement unit
LTI	linear time-invariant
LTP	linear time-periodic
MEMS	micro electro-mechanical system
ODE	ordinary differential equation
OE	output-error
PSE	predicted square error
QUEST	quaternion estimator
SIDPAC	system identification programs for aircraft
TRIAD	tri-axial attitude determination system
UAV	unmanned air vehicle

Preface

"Ornithopters are mechanical contraptions that flap their wings like birds to generate aerodynamic forces... and they don't fly" was my uncharacteristically bold and naively undergraduate response to my (unbeknownst to me) future Ph.D. advisor's question as to whether anyone on my side of the video conference knew anything about ornithopters. He proceeded to tell me that in the time since da Vinci's drawings, ornithopters have been miniaturized and they do now, in fact, fly. Futhermore, he mentioned that he was aquiring a young flock of these birds and was eager to fly, and also that I was to be part of the research effort. And thus were sown the seeds for what took root and grew into an frustratingly enjoyable graduate school experience.

In addition to being a really cool project that liberated me from office walls on sunny days for flight testing, it turns out that ornithopters represent a good middle ground for designing small unmanned vehicles. As aircraft are made smaller, the air appears more viscous due to Reynolds effects, and so flapping wings and unsteady aerodynamics are needed to recover aerodynamic performance. However, the ability to carry sensor payloads diminish and sensitivity to gusts increase due to lower mass and inertia. A miniature ornithopter like that discussed in this monograph is well positioned to perform efficiently, carry payload, and fly in windy environments, as well as achieve agile and maneuverable flight, operate safely in close proximity to humans (despite buzzing videographers during flight tests), and don a degree of contextual camoflauge that has even fooled/attracted/scared other birds. Hence these ornithopters show great promise in aiding people in numerous aspects of life, from crop surveying to terrain mapping, and from military survellience to search and rescue missions.

The material presented herein has stemmed from my dissertation work, which occurred between 2005 and 2011. Needing periodic changes in scenery, I completed parts of this research were conducted at the University of Maryland, the National Institute of Aerospace, and the NASA Langley Research Center, but not in that order and not in one sitting. The original intent of the research was to develop wing morphing technologies using smart structures to actively change the wings in flight and to achieve agility on par with that of birds, or at least to land in a more controlled manner warranting an official label other than "munition." However, flight tests soon indicated that a stabilized flight platform was needed before an actively morphing wing could be implemented. However, flight tests soon indicated that new state estimation and control methods based on new flight dynamics models were needed before a stabilized flight platform could be achieved. Therefore, this book is devoted to developing a new model for ornithopter flight dynamics that is amenable to such tasks in dynamics as simulation, state estimation, system identification, and feedback control.

This book was written with clasped hands and slightly bent knees in the hopes of lifting others onto modest shoulders to see slightly further. The begining is fairly nontechnical: a description of the state of the art, some characterizations of the ornithopter platform, and the (difficult) differences between flapping-wing flight and conventional flight, all of which can be useful to a newcomer in the field or someone with a curious interest. However the pace soon quickens as nonlinear multibody dynamics are introduced and used to derive vehicle dynamics in a form useful for simulation, system identification, state estimation, and feedback control. As with any time the big picture comes finally into focus, I found the classical dynamics presented herin to be epiphanically clear, and I have tried to unify conventions from several references to present the material as I would have liked to learn it, in the hopes of reaching others in such related problems as multibody dynamics, rotor dynamics, aeroelasticity, and control of Euler-Lagrange systems. After the vehicle dynamics were derived, it remained to determine accurate and useful aerodynamic models. It is an understatement to say that this is difficult. Much of the literature has been devoted to analytic theory or conventional wind tunnel testing. Our approach was to combine our nonlinear multibody vehicle dynamics model, wind tunnel data, and flight test data with system identification methods to extract these methods. Completing that, the remainder of the manuscript examines the flight dynamics model, and investigates methods for trimming and linearizing the model, in order to complete that first goal of stabilizing the flight.

As with any journey that spans vast distances in both time and space, the list of those to whom I am thankfully indebted is unweildingly long. In this section I try to acknowledge those who have made the greatest impact on this work. Firstly is my family. My mother, Barbara Grauer, has fed the both the soul and the stomach their respective foods to endure. My father, Lawrence Grauer, in compliment has taught me the skills I have employeed to survive. My sisters Kathryn and Jennifer Grauer have given me a comfortable place to rest. My grandparents Judith Berry and Joyce Spittel have each given me outlets for my frustration. Aunts, uncles, cousins, and the like have all cheered me on. Secondly, the Morpheus Laboratory at the University of Maryland has been a warm center for creativity. Led by Dr. James Hubard, Jr., members of "Gen 1" (Nelson Guerreiro, Robyn Harmon, Benjamin Nickless, Geoff Slipher, and Sandra Ugrina) and "Gen 2" (Cornelia Altenbuchner, Alex Brown, and Aimy Wissa) have comprised what I can truly only describe as a wolf pack of a research group. Many thanks go to Dr. Sean Humbert for supplying equipement and discussions necessary for this work. Dr. Robert Sanner and Dr. Eugene Morelli additionally provided several long discussions that shaped the work. Wind tunnel tests were conducted with the help of Dr. Allan Winklemann, and flight tests were aided by Michael Chinn, Qina Diao, Imraan Faruque, Sarah Haack, Evan Ulrich, and Kan Yang.

Jared Grauer
Hampton, Virginia
May 2012

About the authors

Dr. Grauer earned from the University of Maryland a bachelor's degree in 2005, a master's degree in 2007, and a Ph.D. in 2012. His graduate research experience included teaching, building miniature avionics packages, flying ornithopters, performing system identification on small unmanned air vehicles, and applying feedback control methods. He is currently a research aerospace engineer within the Dynamic Systems and Control Branch at the NASA Langley Research center, where he specializes in expanding system identification and feedback control techniques for new and exotic aircraft.

Dr. Hubbard began his career in 1971 as an engineering officer in the U.S. Merchant Marine serving in Vietnam. He received unlimited horsepower, steam and diesel engine Marine Engineering license from the U.S. Coast Guard and at the age of 19 was one of the youngest to get such an honor. He has served as a Professor of engineering at M.I.T., Boston University and is presently the Langley Distinguished Professor of Aerospace Engineering at the University of Maryland. His research involves the design, analysis, simulation and fabrication of spatially distributed systems, smart materials and smart transducers. He has co-founded three companies whose commercial product base built upon his experience in smart materials and smart sensors. He has received numerous awards for teaching and mentoring excellence including the M.I.T. Goodwin Medal for "Conspicuously Effective Teaching," The M.I.T. Stewart Award for "Outstanding Service to the Community," and in 2002 was awarded "The Key to the City" of his hometown, Danville Virginia for Lifetime achievement and mentoring. He is a Fellow of the AIAA and has more than 100 technical publications, 24 U.S. and Worldwide patents, and has served on numerous technical Boards and Committees of the National Academy of Engineering. He currently resides in Hampton Virginia with his 3 sons, James, Jordan, and Drew, and his wife Adrienne.

1

Introduction

DOI: 10.1533/9780857094674.1

Abstract: This chapter introduces the reader to how ornithopters currently reside in our world, from where they were conceived, and where they are headed. First, the current design market for unmanned air vehicles is discussed, followed by the niche that ornithopters fill. Then current knowledge on the design, aerodynamics, vehicle dynamics, and feedback control is summarized. The chapter ends with defining a path for the contributions presented in this manuscript.

Key Words: state of the art, scope, contributions

1.1 Background and motivation

Unmanned air vehicles (UAVs), or aircraft without on-board pilots, are proliferating into many sectors of society and have the potential to significantly impact daily life. Enabled by advances in electronics miniaturization and composite material manufacturing, hobbyists, commercial companies, defense contractors, universities, and government research laboratories are currently developing small unmanned platforms. UAVs will play an important role in the future because in addition to the relatively low cost, they can un-man aircraft to perform "dull, dirty, and dangerous" work.

In the civilian sector, UAVs are currently used as toys, hobby aircraft, and research flight platforms. They have been employed in such tasks as airport wildlife control and population monitoring, and are expected to contribute to autonomous crop surveying, atmospheric weather monitoring, and search and rescue missions. In the military sector, these vehicles are employed in intelligence, surveillance, and reconnaissance missions. It is envisioned that these vehicles will perform missions such as chemical substance detection and autonomous perimeter surveillance, as well as achieve multi-mission capabilities including perch and stare, robust long-duration outdoor flight, and agile flight through cluttered indoor environments.

To accomplish such mission profiles autonomously, a flight control system, such as that shown Figure 1.1, must be implemented in order to stabilize the vehicle and achieve desired performance goals while attenuating the effects of noise, disturbances, and model uncertainty. The aircraft plant model contains a set of equations that evolve the vehicle position and velocity state variables as a function of the control inputs and disturbances. Sensors on board the aircraft provide measurements, which are used in an estimator to predict the aircraft state. The tracking error is supplied to the controller, which is designed to shape the inputs to achieve flight performance metrics.

Fixed-wing aircraft remain the most prevalent type of UAV, as they are in general relatively simple aircraft and the dynamic models describing their motions are well known and understood [1, 2, 3]. However, as the vehicle size decreases and viscous effects become more pronounced, fixed-wing aircraft suffer from decreased lift to drag ratios that degrade flight performance [4]. Additionally, most fixed-wing aircraft require forward speed to generate aerodynamic forces, making it difficult for them to hover and fly slowly, which are abilities needed for indoor flight.

Miniature rotary-wing vehicles present an alternative to fixed-wing designs. These vehicles are highly maneuverable and have the ability to hover, making them suitable for indoor reconnaissance and surveillance missions; however, rotorcraft have a number of drawbacks. As with fixed-wing aircraft, viscous effects reduce the aerodynamic efficiency of the vehicle [5]. The added complexity of the rotor and swash plate systems increases the vehicle cost and maintenance.

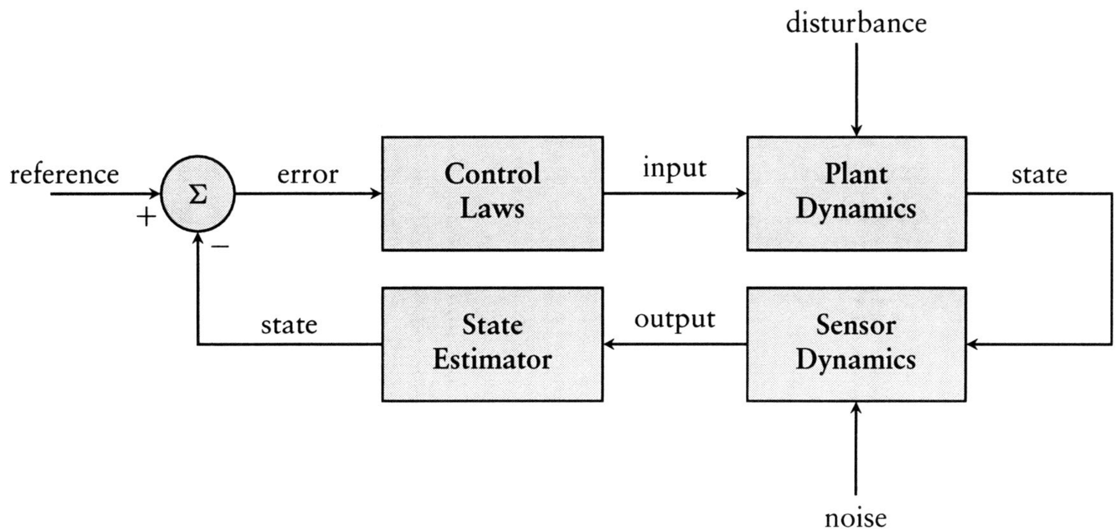

Figure 1.1 Flight control architecture block diagram

Furthermore, these aircraft are noisy, open-loop unstable, and potentially hazardous to bystanders and operators, making them difficult for human operators to fly without training and closed-loop control.

Although both conventional fixed-wing and rotary-wing vehicles fill niches in the design space of UAVs, a new vehicle design is needed to both fly in a robust manner outdoors and in an agile manner in cluttered environments.

1.2 Bio-inspired flapping wing aircraft

Miniature air vehicles that use flapping wings to generate aerodynamic forces and moments have advantages over both fixed- and rotary-wing vehicles. For vehicle sizes on the order of 0.1 kg and below, flapping wings are required to gain maneuverability and aerodynamic efficiency [6, 7, 8]. These vehicles are also safe for humans to operate, as they consist only of light weight material and do not have spinning rotors or fuel tanks, and they don a level of contextual camouflage due to the bird-like appearance. It is envisioned that by mimicking the agility, maneuverability, and robustness of natural flyers, such as insects, bats, and birds, these vehicles will fill the demand for multi-mission capable UAVs.

Avian flapping flight is characterized by the up stroke and the down stroke of the wings. A cross-section of a seagull wing in forward flight is illustrated in Figure 1.2, as viewed from a body-fixed reference frame [9]. Lift and drag forces are generated over the wing, perpendicular and parallel to the local flow direction, respectively, due to the forward motion of the vehicle and the plunging motion of the wing. During the down stroke, the resultant force is pointed such that a propulsive thrust force and a lift force are transmitted to the body, whereas the resultant force is oriented such that a lift force and a drag force are transmitted to the body during the up stroke. Typically the wing generates most of the lift force near the root of the wing, whereas the majority of the

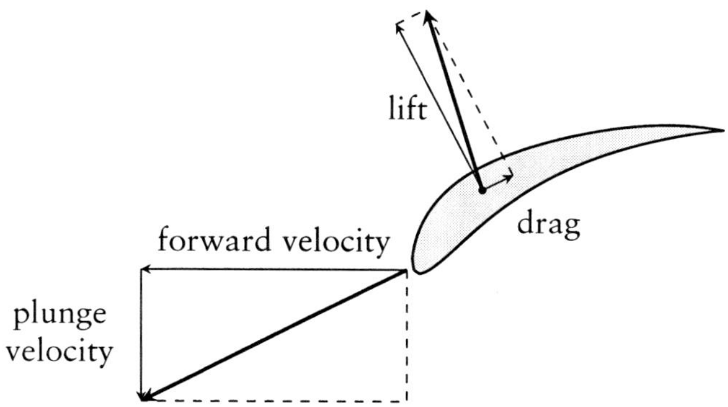

(a) down stroke

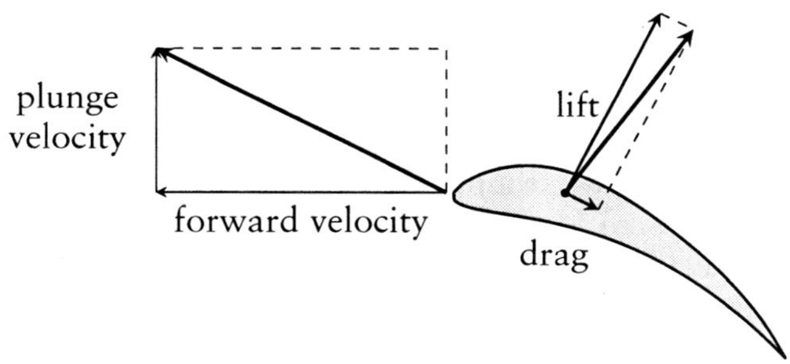

(b) up stroke

Figure 1.2 **Forces acting on an avian wing section, as viewed from a body-fixed reference frame**

thrust is created on the outboard sections of the wing. Asymmetries between the up stroke and down stroke are required to produce a net lift and thrust force over the flapping cycle. Natural fliers achieve this effect either by flying with a positive mean angle of attack, modulating the up/down stroke ratio, or by changing the wing kinematics through wing morphing [10, 11].

1.3 Flapping-wing literature review

1.3.1 Vehicle design

The first recorded ornithopter vehicle design was da Vinci's 1490 concept for a human-powered ornithopter, having membrane wings and a mechanically efficient transmission. In Germany during 1894, Lilienthal created the "Kleiner Schlagflügelapparat," a human powered flapping device inspired by his previous glider work and experiments with birds. During the late 19th century designers began experimenting with new forms of power generation, including Trouvé's 1870 flight powered by gunpowder charges and Hargrave's steam-powered ornithopters in 1890. Beginning in 1942 with Schmid's design, ornithopters were powered using internal combustion engines. Man-powered ornithopter flights extending the vehicle endurance time were achieved by Lippisch, Maule, Hartmann, Toporov, and Rousseau between 1929 and 1993. The most successful manned ornithopter has been that of the University of Toronto, which is able to take off under its own power [12].

Unmanned ornithopter designs began with the rubber band powered models developed by Jobert, Pénaud, and Villeneuve in the 1870s. Begining with Spence's 1961 ornithopter, designers incorporated smaller power sources and radio electronics into the vehicles. Similar ornithopters include the quarter scale model by DeLaurier in 1991 and the "Elektro Vogel" series of ornithopter started by Räbinger in 1975 [13]. Ornithopters for hobbyists have also developed, including the Kinkade and Cybird ornithopters, in the late 1990's. Universities have designed ornithopters, including the University of Arizona ornithopter [14], "Microbat" at the California Institute of Technology [15], the "Phoenix" ornithopter at the Massachusetts Institute of Technology [16], and the Morpheus ornithopter at the University of Maryland [17].

1.3.2 Aerodynamics modeling

It is difficult to model the aerodynamics of a flapping wing. One problem is that the Reynolds number for these vehicles is typically under 10^5, which encompasses a regime where aerodynamic models are currently innaccurate. Additionally, the thin membrane airfoils and light weight structures add complex aero-elastic couplings which require additional states or finite-element models to capture. Furthermore, the flapping motion of the wings generates complex, unsteady, three-dimensional flow fields that are difficult to model.

It is generally accepted in the literature that the aerodynamic lift force of flapping wings can be crudely, but analytically, represented using the quasi-steady model [18]

$$C_L = C_{L_\alpha}\alpha + C_{L_{\dot{\alpha}}}\dot{\alpha} + C_{L_{\dot{h}}}\dot{h} \qquad (1.1)$$

where the lift coefficient is a function of the instantaneous angle of attack α, its time derivative $\dot{\alpha}$, and the local plunge velocity $\dot{h}$. The drag coefficient is often modeled using the drag polar

$$C_D = C_{D_0} + C_{D_\alpha}\alpha + C_{D_{\alpha^2}}\alpha^2 \qquad (1.2)$$

in order to capture the parasite drag and drag parabola. Often times these aerodynamic models are posed using two dimensional airfoil sections in a blade element model [18, 19, 20, 21, 22, 23], where the local flow velocities may be more accurately represented. Coefficients in these models are derived from first principles or measured from experiments.

The most popular analytical model for avian flyers is the model developed by DeLaurier, who used a quasi-steady blade element model to capture wake/vortex interactions, post-stall phenomenon, and partial leading edge suction. Variants of this model are used widely in the literature [24, 25, 26, 19, 20, 27, 28].

Several attempts to empirically measure aerodynamic models from data have also been conducted, mostly as wind tunnel tests where the ornithopter is mounted to a load cell measuring lift and drag forces, resulting in a look-up table of aerodynamic forces and moments [14, 29, 27]. This method constrains the ornithopter in translation and rotation, thus altering the natural response of the vehicle and aerodynamic flow environment seen by the wings, and is therefore not representative of free flight aerodynamics.

Numerical tools have also been developed to examine flapping flight. While these methods require more computation time and can not be used directly for feedback control, they provide insight into the fundamental flight physics, and can be reduced to obtain low-order models. Panel methods solving three dimensional, unsteady, incompressible flows around avian-based flapping models in wind tunnels have shown good results [30]. Reynolds-averaged computational fluid dynamics solvers have also been used to illuminate the aerodynamic flow field around flapping wings and contribute the the fundamental understanding of flapping flight [31, 32].

The field currently lacks an accurate and low-order model for predicting aerodynamic flows over wings, useful in a control-theoretic framework. This is in part due to the variety of ornithopter designs, having membrane wings, thin airfoils, and structural deformations. Another reason is the lack of experimental data available for flapping-wing vehicles in free flight, due in part to the difficulty in obtaining measurements. Recently developed conformal sensors show potential in measuring lift and drag forces in real-time and produce direct measurements of the aerodynamic flow fields [33]. Additionally, non-invasive, vision-based measurements are becoming popular, which when combined with system identification methods can produce aerodynamic measurements.

1.3.3 Vehicle dynamics modeling

There are several types of models that describe the flight dynamics of flapping-wing aircraft. These models are traditionally derived first as nonlinear models that evolve position and velocity states. For larger aircraft that have significant mass and inertia in the wings, and/or where the flapping frequencies are near the rigid body and control frequencies, multibody models are required. These models are more complicated and contain more states, but can capture the inertial effects of the flapping wings. Orlowski concluded, based upon a multibody model and empirical scaling laws observed in nature, that linear momentum effects from flapping wings are always significant, but angular momentum effects begin to attenuate for flapping frequencies above 40 Hz [23]. Rigid body dynamics for multibody ornithopters have been derived using a variety of methods [21, 34, 35, 36], and have also been included with finite-element models to capture fluid/structure interactions [28].

If the inertial effects of wings are negligible and the wings flap significantly faster than the dynamics of interest, then the conventional single-body equations of motion for aircraft [1, 37, 2] can be used to model the vehicle dynamics. These models are simpler, and the wing motion is either a kinematic state or a control input. This approach is common in the literature [38, 39, 40], even for larger vehicles where its validity is questionable [25, 14, 41].

Linear perturbation models are often desired for stability analysis and controls synthesis. Linear time-invariant (LTI) models result in equilibrium points with Eigenvalues and Eigenvectors, for which a large set of classical and modern control tools can be applied. Several authors have employed averaging theory to simplify the periodic forcing of the wings [40, 42, 23]. Taylor et al. computed the state matrix using empirical finite differences on insects in a wind tunnel [43, 44]. Faruque and Humbert used system identification to obtain a low-order equivalent system for an insect [45, 46]. Kim et al. used a linear discrete-time model of hyper-states to map inputs to outputs [47]. Analyses have resulted in a wide variety of modal structures, most of which have unstable motions. Several authors have questioned the validity of these LTI models [43, 44, 47, 48] and have consequently developed more complex models [49].

The higher fidelity models result in trim conditions characterized as limit cycle oscillations in the state space, rather than fixed points. Stability has been inferred from simulation runs [49], and also proven analytically. For instance, Dietl and Garcia applied Floquet analysis using Poincaré maps to create an analogous discrete-time system. Both their findings and a similar study by Bolender showed unstable trim solutions [25, 50].

1.3.4 Feedback control

One of the goals of modeling the flapping wing vehicle is to be able to control the vehicle and complete mission scenarios autonomously. These goals range from robust way-point navigation in outdoor gusty environments, to agile flight through cluttered indoor environments, to outdoor precision perching maneuvers.

Smaller flapping wing vehicles flap at frequencies too high to perform continuous control; instead, a control signal is computed at the beginning of each flap cycle, in a sample and hold fashion. Several authors used variables parameterizing the stroke kinematics of each wing, as well as a mechanical bob weight, to achieve a fully actuated system capable of position tracking in simulation [22, 26, 38]. Deng et al. used averaging theory to obtain an analogous discrete-time, under-actuated system, for which stabilized hover was demonstrated in simulation [42].

Control methods for larger vehicles resemble those of standard aircraft. Krashanitsa et al. integrated a commercial fixed-wing autopilot into an ornithopter and was able to demonstrate navigation between two way point coordinates using tuned proportional control loops [14]. The work demonstrated successful closed-loop control, but also the limits of using conventional avionics and control without a model for flapping-wing aircraft. Recently, Dietl and Garcia explored shortcomings of using classical control and suggested a discrete-time linear quadratic regulator control based on a linear time-periodic model [51]. Lee et al. has also recently explored using a bio-inspired nonlinear control for the tail to attenuate pitch oscillations and stabilize the vehicle [28]. Tedrake has suggested coupling a simple flight dynamics model with an advanced machine learning technique to achieve agile flight [52].

1.4 Scope and contributions of current research

There is currently no flight dynamics model of ornithopter flight that is both accurate and amenable for control purposes. The goal of this work is to create such a model, with insights into design and control, so that the community may progress towards realizing an autonomous ornithopter flight platform.

In Chapter 2, standard nomenclature is discussed and the experimental flight platform used in this study is introduced. As models and flight data for this type of vehicle are not available in the literature, a series of experimental investigations into the characteristics of flight are presented. The results suggest a nonlinear multibody model is needed to capture the dynamics. Those nonlinear multibody vehicle dynamics are derived in Chapter 3. Energy methods are used to determine equations of motion, which are then cast into a canonical form that is useful for simulation, system identification, and nonlinear feedback control design.

Chapter 4 presents the system identification work performed to identify parts of the ornithopter flight dynamics model that are not known a priori and are difficult to model analytically. A wind tunnel test is discussed where lift and drag models for the ornithopter tail were identified. Afterwards, an experiment is presented using a visual tracking system to obtain flight test data, from which an aerodynamic model for the wings was identified.

Chapter 5 presents flight simulation results obtained using the flight dynamics model developed in Chapters 3 and 4. A method for determining trim solutions is presented and is employed to trim the ornithopter model for straight and level mean flight. The model is then linearized about the trim trajectory, resulting in a canonical time-invariant model as well as a time-periodic model. Feedback control laws are then designed for the ornithopter longitudinal dynamics that could be implemented in an autopilot.

The original contributions of this research to the state of the art are the following:

- Flight data suitable for modeling
  This work presents flight data of an ornithopter, sampled at a bandwidth that illuminated the effects of the flapping wings. Trimmed flight data showed pitch oscillations up to 4.97 rad/s and heave accelerations up to 41.7 m/s^2, which require nonlinear multibody models to capture. Additionally, harmonic responses were seen in the data, indicating either structural modes or nonlinearities. High accelerations preclude the use of traditional attitude determination methods and require model-based techniques for state estimation.
- Modeling of mass distribution variation due to flapping wings
  The effects of the flapping motion on the mass distribution of the ornithopter were investigated using a CAD model. The center of mass travels 10.3 cm in the vertical direction, the moments of inertia can vary up to 53.6%, and the inertia rates are significant. Multibody models are required to capture these variations.
- Derivation of rigid multibody vehicle dynamics
  It was determined that a three-body model was sufficient to model the ornithopter. Vehicle dynamics were derived using the Boltzmann-Hamel equations, and were cast into a canonical form used for the nonlinear control of Euler-Lagrange systems.
- Tail aerodynamics system identification
  Wind tunnel tests were conducted on the ornithopter tail, having free stream velocities, angles of attack, and angles of sideslip ranging between 4.50 to 6.33 m/s, −0.88 to −0.35 rad, and ±0.57 rad, respectfully. System identification methods were applied to determine models of the aerodynamic coefficients.
- Flight testing and wing aerodynamics system identification
  Flight data, obtained using a visual tracking system, was presented. This data and the vehicle dynamics model were used to explore variations in angle of attack, Reynolds

number, reduced frequency, velocity, and structural deformations over the wing and throughout the wing stroke cycle. Additionally, force and moment sources were identified throughout the wing stroke. This data was also used to determine aerodynamic models for the wings using system identification techniques.

- Models for stability and control
  In addition to the full nonlinear model, a linear time-invariant model and a linear time-periodic model are presented. The time-invariant model is shown to not contain enough accuracy for control design; rather, the time-periodic model is needed.

2

Ornithopter test platform characterizations

DOI: 10.1533/9780857094674.15

Abstract: The form and structure of mathematical models describing conventional aircraft flight dynamics are well known; however, it is not obvious whether these same models can be used to describe the flight dynamics of ornithopters. This chapter presents experimental investigations into flapping-wing flight using an ornithopter test platform. Standard nomenclature for describing aircraft flight dynamics is introduced, followed by a description of the ornithopter test platform. Afterwards, flight data is presented that exhibits nonlinear phenomenon including fast and large amplitude motions, as well as limit cycle oscillations. The mass distribution of the ornithopter is then explored, and it is shown that the flapping wings induce significant changes in the center of mass location, inertia tensor, and inertia rates. An experiment is summarized where measured lift and thrust forces were modeled using quasi-steady aerodynamics. Finally, implications for flight dynamics modeling are discussed and a new model for flapping-wing aircraft is outlined.

Key Words: aircraft dynamics, flight data, mass distribution

2.1 Mathematical representation of an aircraft

Consider the generic aircraft shown in Figure 2.1. An inertial reference frame $K_I = \{\mathbf{e}_{xI}, \mathbf{e}_{yI}, \mathbf{e}_{zI}\}$ is fixed on the surface of the Earth at C_I with unit vectors pointing North, East, and down, respectively. A body frame $K_0 = \{\mathbf{e}_{x0}, \mathbf{e}_{y0}, \mathbf{e}_{z0}\}$ is fixed to the aircraft center of mass C_0, located at

$$\mathbf{r}_{0,I}^{I} = \begin{bmatrix} x \\ y \\ z \end{bmatrix}, \tag{2.1}$$

with axes pointing out the nose, right wing, and underside of the aircraft. A wind frame K_W is collocated with the body frame and is oriented into the oncoming wind.

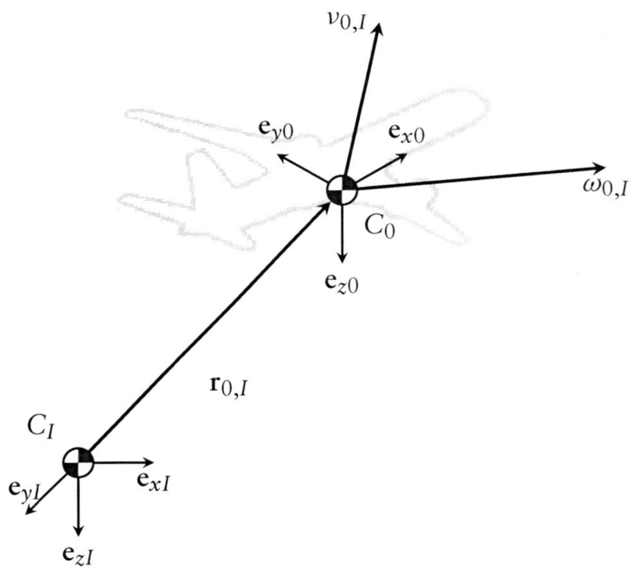

Figure 2.1 An annotated generic aircraft

The attitude, or orientation, of the vehicle is specified by $\eta^I_{0,I}$. An Euler angle parametrization admits the three-element attitude vector

$$\boldsymbol{\eta}^I_{0,I} = \begin{bmatrix} \phi \\ \theta \\ \psi \end{bmatrix} \tag{2.2}$$

with corresponding rotation matrix

$$\begin{aligned} \mathbf{R}^{0,I} &= \begin{bmatrix} 1 & 0 & 0 \\ 0 & \cos\phi & \sin\phi \\ 0 & -\sin\phi & \cos\phi \end{bmatrix} \begin{bmatrix} \cos\theta & 0 & -\sin\theta \\ 0 & 1 & 0 \\ \sin\theta & 0 & \cos\theta \end{bmatrix} \\ &\times \begin{bmatrix} \cos\psi & \sin\psi & 0 \\ -\sin\psi & \cos\psi & 0 \\ 0 & 0 & 1 \end{bmatrix} \end{aligned} \tag{2.3}$$

where ϕ, θ, and ψ are the roll, pitch, and yaw angles. Alternatively, a quaternion parametrization admits the four-element attitude vector

$$\boldsymbol{\eta}^I_{0,I} = \begin{bmatrix} \epsilon \\ \delta \end{bmatrix} \tag{2.4}$$

and corresponding rotation matrix

$$\mathbf{R}^{0,I} = (\delta^2 - \epsilon^T\epsilon) + 2\epsilon\epsilon^T - 2\delta S(\epsilon) \tag{2.5}$$

where ϵ is the vector part and δ is the scalar part of the quaternion, and where the skew operator is defined

$$S(x) = \begin{bmatrix} 0 & -x_3 & x_2 \\ x_3 & 0 & -x_1 \\ -x_2 & x_1 & 0 \end{bmatrix} \tag{2.6}$$

for any three-element vector.

The aircraft has body-fixed translational velocity

$$\boldsymbol{\nu}^0_{0,I} = \begin{bmatrix} u \\ v \\ w \end{bmatrix} \tag{2.7}$$

and body-fixed rotational velocity

$$\boldsymbol{\omega}_{0,I}^{0} = \begin{bmatrix} p \\ q \\ r \end{bmatrix}. \tag{2.8}$$

Finally, the local airspeed, angle of attack, and sideslip angle

$$\begin{aligned} V &= \sqrt{u^2 + v^2 + w^2} \\ \alpha &= \arctan(w/u) \\ \beta &= \arcsin(v/V) \end{aligned} \tag{2.9}$$

are used to construct the rotation matrix

$$\mathbf{R}^{W,0} = \begin{bmatrix} \cos\alpha\cos\beta & \sin\beta & \sin\alpha\cos\beta \\ -\cos\alpha\sin\beta & \cos\beta & -\sin\alpha\sin\beta \\ -\sin\alpha & 0 & \cos\alpha \end{bmatrix} \tag{2.10}$$

to rotate from the body frame to the wind frame.

2.2 Ornithopter aircraft description

The commercially available "Slow Hawk" ornithopter, by Sean Kinkade [53], was selected for this study due to its payload capacity, durability, and handling characteristics. A photograph of this ornithopter is shown in Figure 2.2 and relevant aircraft parameters are provided in Table 2.1.

The fuselage is a 3 mm ply of carbon fiber and has affixed the electronic components. The wings are a rip-stop polyester membrane having 5 mm and 3 mm diameter carbon fiber spars along the leading edge and near the trailing edge, respectively. The tail is also a membrane, with small carbon fiber spars running from forward to aft. Planform geometries are illustrated in Figure 2.3.

To fly the ornithopter, the aircraft is launched into the air and then piloted using standard 72 MHz hobby radio electronics. Power is delivered from a two-cell, 7.4 V, 1320 mAh lithium-polymer battery to an electronic speed controller, which drives a brushless DC motor. This motor is attached to

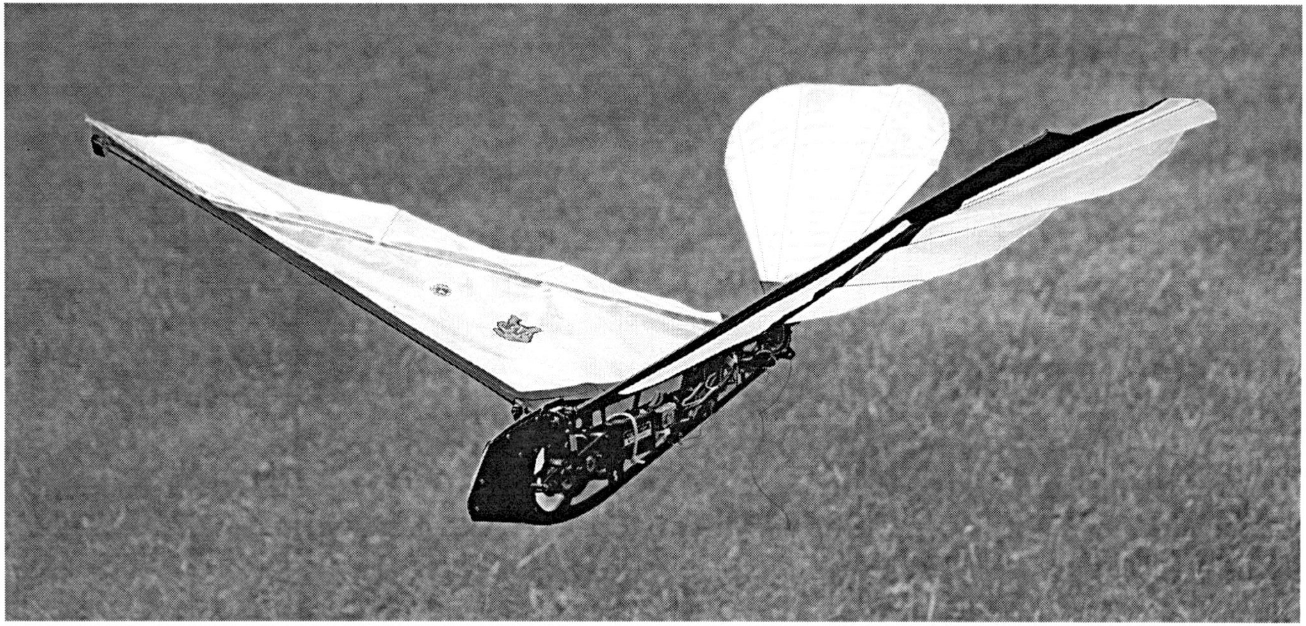

Figure 2.2 Ornithopter flight test platform

Table 2.1 Aircraft parameters

Parameter	Symbol	Value	Unit
total mass	m	0.45	kg
wing span	b_w	1.22	m
wing mean aerodynamic chord	$\bar{c}_w$	0.29	m
wing area	S_w	0.30	m^2
wing aspect ratio	AR_w	4.40	-
tail span	b_t	0.20	m
tail mean aerodynamic chord	$\bar{c}_t$	0.20	m
tail area	S_t	0.04	m^2
tail aspect ratio	AR_t	1.50	-

a gear box, four-bar mechanism, and leading edge wing spar, and drives the flapping motion of the wing. As the throttle is increased, the wings flap faster, thereby increasing the lift and thrust. Power is also delivered to two servo motors which pitch and roll the tail in a serial fashion to create pitching and yawing moments on the aircraft, respectively. Flight duration for this aircraft is approximately 15 minutes.

2.3 Measurements from flight data

Flight data can help guide the flight dynamics modeling process; however, there was no adequate flight data for ornithopter flight available in the literature. As such, the ornithopter was outfitted with an avionics unit to record vehicle pose and inertial sensor data during flight. This section presents the experimental setup, flight test, and data analysis. See reference [54] for a complete discussion.

2.3.1 Experimental setup

Autopilot and stability augmentation systems commonly used on aircraft employ sensors including magnetometers, accelerometers, and gyroscopes to provide information on the vehicle motion. Commercially available packages for

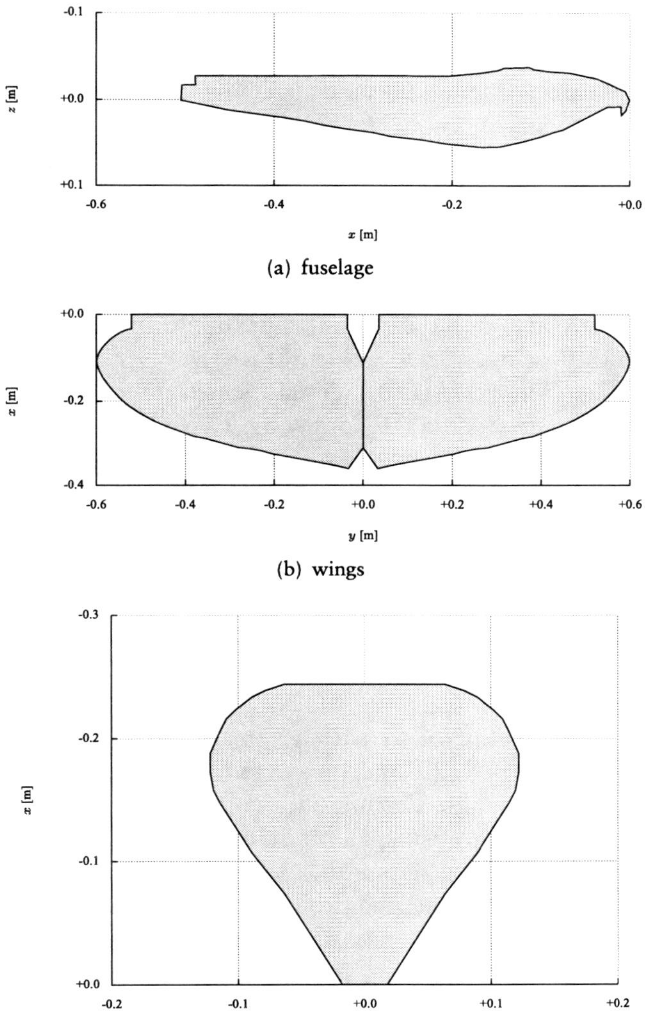

(a) fuselage

(b) wings

(c) tail

Figure 2.3 Ornithopter planform geometries

unmanned aircraft typically include sensors that have range and bandwidth too low to investigate ornithopter flight. A custom avionics package was therefore developed to provide measurements in flight from inertial and pose sensors.

The avionics board, pictured in Figure 2.4 has a 40 cm by 70 cm footprint and a 27.9 g mass. The board was designed using CadSoft Eagle, manufactured by an external company, and populated using re-flow solder equipment. Power is supplied by a dedicated 2-cell, 7.4 V, 200 mAh lithium-polymer battery. Four voltage regulators deliver power at 3.3 V, 5 V, and 7 V for the on-board processing and sensors. Additional connections are included to reprogram the microprocessor and to connect to the RS-232 interface

The MemSense MAG3D inertial measurement unit (IMU) was chosen for its small 1.78 cm by 1.78 cm footprint and 10 g mass. This IMU provides analog voltage outputs, filtered to 50 Hz, from orthogonal triads of magnetometers, gyroscopes, and accelerometers, which measure local magnetic fields $\mathbf{h}^0_{0,I}$, body-fixed rotational velocities $\omega^0_{0,I}$, and local accelerations $a^0_{0,I}$, respectively. Additional sensors were installed on the ornithopter to measure temperature, power consumption, wing angle, pilot stick commands, and tail surface deflections.

The board is controlled with a reprogrammable 40 MHz Microchip PIC18F8722 microprocessor, programmed in the C computer language. During data collection, analog sensor outputs are digitized using 16 bit analog to digital converters and are recorded on a removable 1 GB memory card. Aggregate delays from data acquisition, computation, and memory write times resulted in a maximum sampling rate of 146 Hz. Measurement specifications are summarized in Table 2.2.

The IMU was installed on the ornithopter at the wings level center of mass location, which mitigated DC motor magnetic field effects and mechanical vibrations, as well as minimized accelerometer corrections due to rotation [55]. Sensors that measured power consumption, wing angle, pilot stick inputs, and tail surface deflections were calibrated in the laboratory. Inertial sensors were calibrated at the time and place of flight testing using the method presented in Appendix 6.4.

Table 2.2 **Avionics measurement specifications**

Measurement	Range	Resolution	Error	Unit
time	-	0.03×10^{-3}	-	s
throttle input	0 to 1	1.53×10^{-3}	1.42×10^{-3}	-
longitudinal input	-0.87 to -0.31	8.52×10^{-6}	6.85×10^{-3}	rad
lateral input	± 0.70	0.02×10^{-3}	7.44×10^{-3}	rad
wing angle	± 3.14	0.10×10^{-3}	20.8×10^{-3}	rad
longitudinal angle	-0.87 to -0.31	9.37×10^{-6}	25.1×10^{-3}	rad
lateral angle	± 0.70	0.02×10^{-3}	55.1×10^{-3}	rad
magnetic field	± 3.17	0.05×10^{-3}	5.83×10^{-3}	-
rotational velocity	± 5.24	0.08×10^{-3}	0.14×10^{-0}	rad/s
linear acceleration	± 98.1	1.50×10^{-3}	49.6×10^{-3}	m/s^2

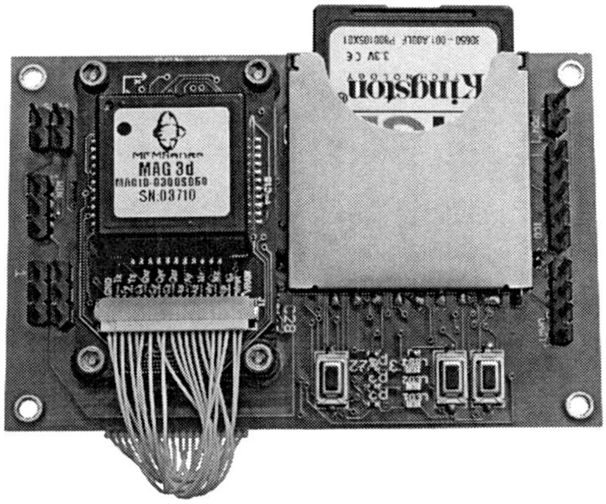

(a) obverse

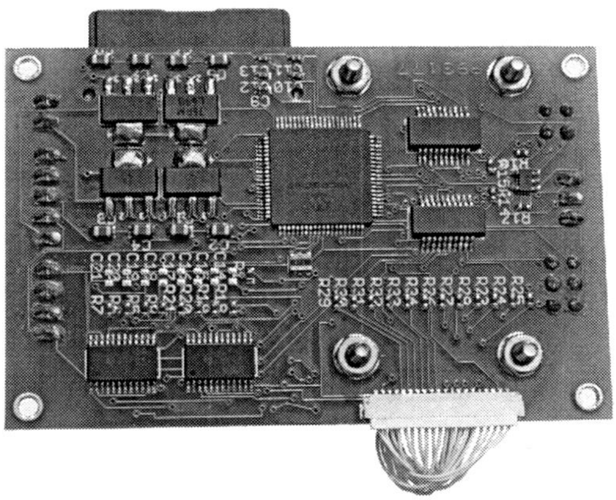

(b) reverse

Figure 2.4 **Custom avionics board used to record flight data**

2.3.2 *Results*

A flight test was conducted where the ornithopter was manually piloted and trimmed for straight and level mean flight. A 6.4 s segment of data containing 28 complete wing strokes was chosen for analysis; a confined flight volume and the presence of wind gusts prohibited longer segments of usable flight data.

The flight data is periodic with the flapping period $\mathcal{T}_f$. As such, ensemble averaging techniques were applied to compute mean wave forms and variances. For a noisy measurement of a $\mathcal{T}_f$-periodic signal y_m, the ensemble average and variance are

$$\check{y}_m(t) = \frac{1}{n}\sum_{k=0}^{n-1} y_m(t + k\mathcal{T}_f) \tag{2.11}$$

$$\sigma_y^2(t) = \frac{1}{n}\sum_{k=0}^{n-1} \left[y_m(t + k\mathcal{T}_f) - \check{y}_m(t)\right]^2 \tag{2.12}$$

where here n is the number of complete periods contained within the data.

Ensemble averages are shown in Figure 2.5 with 2σ error bounds. The wing angle δ_w, defined as the roll angle of the right wing relative to the fuselage, oscillates with a 0.44 rad amplitude at 4.69 Hz. Magnetometer measurements, normalized to the strength of the magnetic field of the Earth, exhibit oscillations in the pitch axis. The gyroscopes show that while there is some roll and yaw oscillation, the dominant motion is in the pitch rate, which obtains a maximum value of $+4.97$ rad/s at $0.40\mathcal{T}_f$ and a minimum value of -4.36 rad/s at $0.90\mathcal{T}_f$. Periodic forcing of the wings and gravitational forces create large heave motions, which are evident by the $+41.7$ m/s^2 acceleration at $0.25\mathcal{T}_f$, and the -21.8 m/s^2 acceleration at $0.68\mathcal{T}_f$. These fast and large amplitude motions observed in trimmed flight require nonlinear models to capture. Moreover, the fact that these signals are periodic indicates the presence of limit cycle oscillations, which are a nonlinear phenomenon. Increases in error bounds at $0.75\mathcal{T}_f$ in the rotational velocity and linear acceleration measurements may indicate the presence of unsteady aerodynamic forces in the transition from down stroke to up stroke.

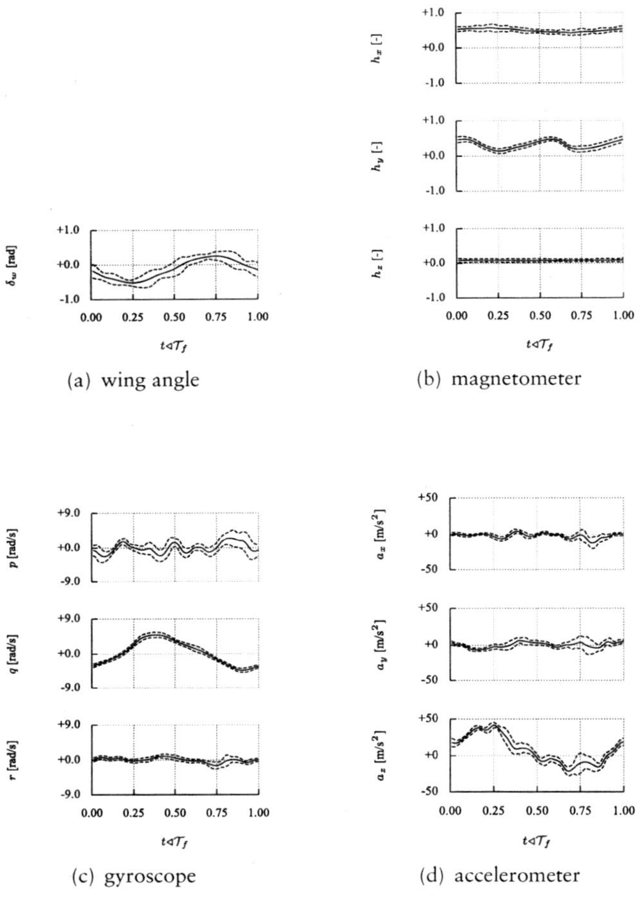

Figure 2.5 **Ensemble averages of flight data with two standard deviation error bounds**

The frequency content of the flight data was also examined using a high accuracy implementation of the chirp-z algorithm [56, 37] to compute the Fourier transform

$$\mathcal{F}\{y(t)\} = \int_0^t y(\tau)e^{-j\omega\tau}d\tau \tag{2.13}$$

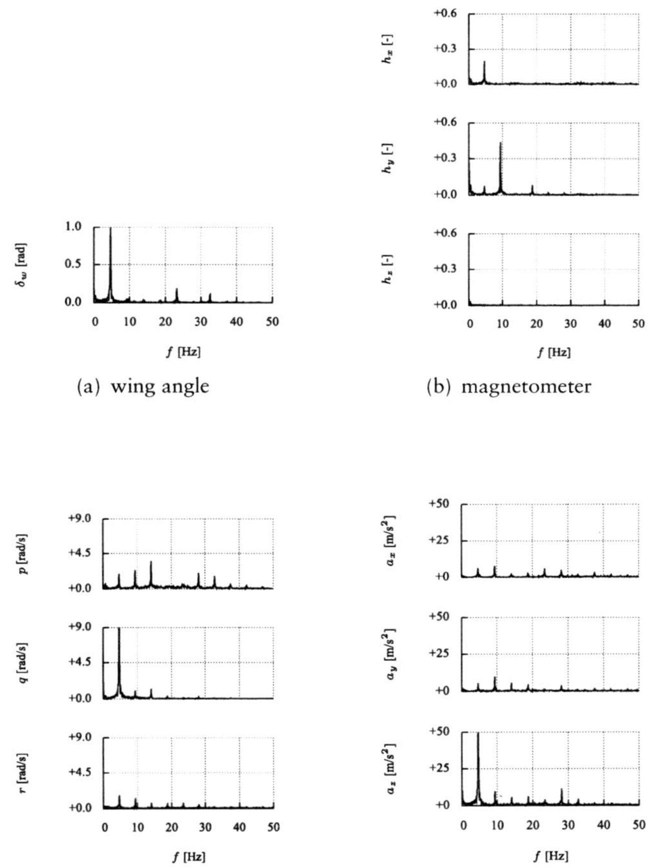

(a) wing angle (b) magnetometer

(c) gyroscope (d) accelerometer

Figure 2.6 **Power spectrum of flight data**

using an arbitrarily specified set of frequencies ω. Frequency transforms were computed using a 0.10 Hz spacing up to 50 Hz, and are shown in Figure 2.6. The sensors show power at harmonics of the flapping frequency, which indicate structural modes in the airframe and/or nonlinear effects in the flight dynamics.

2.4 Configuration-dependent mass distribution

Flight dynamics models typically approximate aircraft as a single rigid body and neglect changes in mass distribution due to rotating machinery, control surface deflections, and fuel burn. The ornithopter wings however have appreciable mass compared with the rest of the vehicle and flap at frequencies near the expected rigid body modes of the system. In this section it is shown that the flapping wings create significant changes in the center of mass location, inertia tensor, and inertia rate, which cannot be captured by the standard aircraft flight dynamics model.

The ornithopter consists of three sections of parts: the fuselage, wings, and tail. For this ornithopter, the moving parts constitute a significant portion of the vehicle mass, as illustrated in Figure 2.7. Both a component build up method and a computer-aided design (CAD) program were employed to model the vehicle mass distribution. The first method consisted of measuring part masses and geometries from which inertia tensors were then computed and transformed to the vehicle center of mass. In the second method, software was used to compute the inertia tensor from mass measurements and three-view photographs. Both methods produced consistent results, however the CAD results were used because they are generally considered to be more accurate.

The effect of wing position on the gross vehicle center of mass location, referenced to the fuselage center of mass, is shown in Figure 2.8(a). The symmetric motion of the wings creates a 10.3 cm (0.36 $\bar{c}_w$) variation in the vertical center of mass position. The vertical bias is due to the wings-level position being located above the fuselage center of mass. This change in center of mass location introduces complexity into the conventional aircraft model, as the body and stability axes are located at the center of mass.

The effect of wing position on the gross vehicle inertia tensor, referenced to the gross vehicle center of mass, is shown in Figure 2.8(b). The products of inertia are small and do

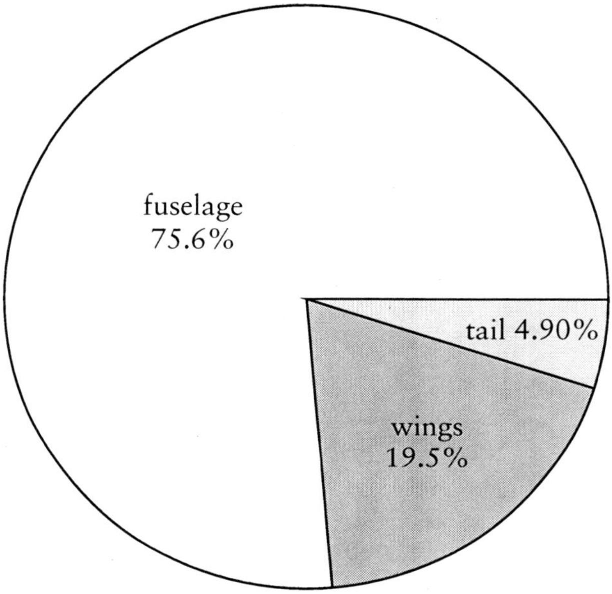

Figure 2.7 **Ornithopter mass distribution contributions**

not exhibit large changes. The roll inertia shows 32.2% variations due to the wing hinge location above the gross center of mass location. The pitch inertia is biased due the hinge offset and the moving center of mass, and experiences a 53.6% increase in value as the wings rotate. The yaw inertia is centered about the wings-level position and experiences a symmetric 52.6% change in inertia as the wings move. These variations introduce a position-dependent inertia matrix into the standard aircraft model.

The time rate of change of the inertia tensor also contributes inertial forces to the equations of motion. Inertia rates resulting from a sinusoidal approximation to the wing trajectory

$$\delta_w = 0.56 \sin(2\pi f_f t) - 0.13 \qquad (2.14)$$

are shown in Figure 2.9 as a function of the wing angle, where the flapping frequency $f_f = 1/T_f$ is varied between 0 and 10 Hz. The pitch inertia rate is smaller in amplitude

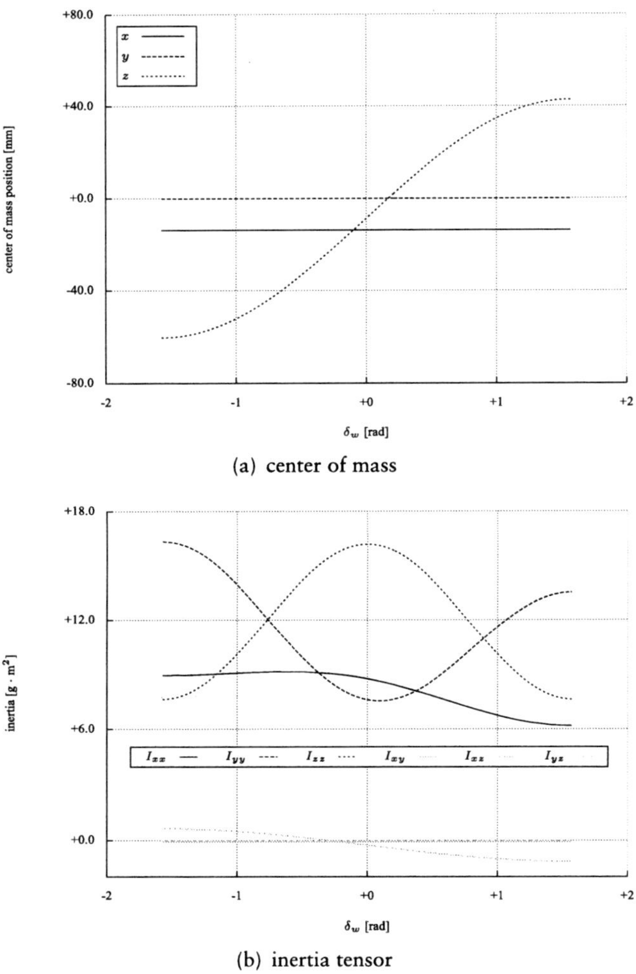

Figure 2.8 **Mass distribution due to wing position variation**

because the pitch inertia experiences smaller changes, and is much larger above −0.5 rad due to a greater slope. The pitch rate inertia and yaw rate inertia are null at approximately 0.1 rad and 0 rad, where the inertias reach a minimum and a maximum, respectively. The plots grow in magnitude for negative wing angles because the asymmetric nature of (2.14) puts more travel and faster velocity in that regime. The inertial rate effects are on the same order as magnitude as the inertial effects for the ornithopter and require multibody models to capture.

2.5 Quasi-hover aerodynamics

It is difficult to model aerodynamic forces and moments on an arbitrary flapping wing aircraft from first principles. To experimentally study the ornithopter aerodynamics, force and wing shape measurements were collected while flapping in a quasi-hover state.

Lift and thrust measurements were recorded from a six-axis strain gauge transducer, and wing angle measurements form a potentiometer, at 1000 Hz using LabVIEW. The ornithopter wing was also fitted with one hundred 3 mm diameter retro-reflective spherical markers. A visual tracking system, consisting of six cameras and a software suite, was used to track the spatial position of each marker at a rate of 350 Hz, providing a discretized measurement of the wing shape. The ornithopter was flapped at a number of frequencies between 3 and 7 Hz while the aerodynamic forces, wing angle, and wing shape were recorded.

A representative segment of lift and thrust data is shown in Figure 2.10, where the ornithopter was flapping at 4.7 Hz. Thrust forces range between 0.26 and 6.86 N, and oscillate at twice the flapping frequency, with the larger peak occurring at the transition between down stroke to up stroke and the smaller peak occurring at the transition between up stroke and down stroke. The lift force oscillates at the flapping frequency, and is zero mean with a 15.6 N amplitude.

This experiment resulted in two models that used wing shape data as input to validate the aerodynamic forces. The first was a computational fluid dynamics model, which used

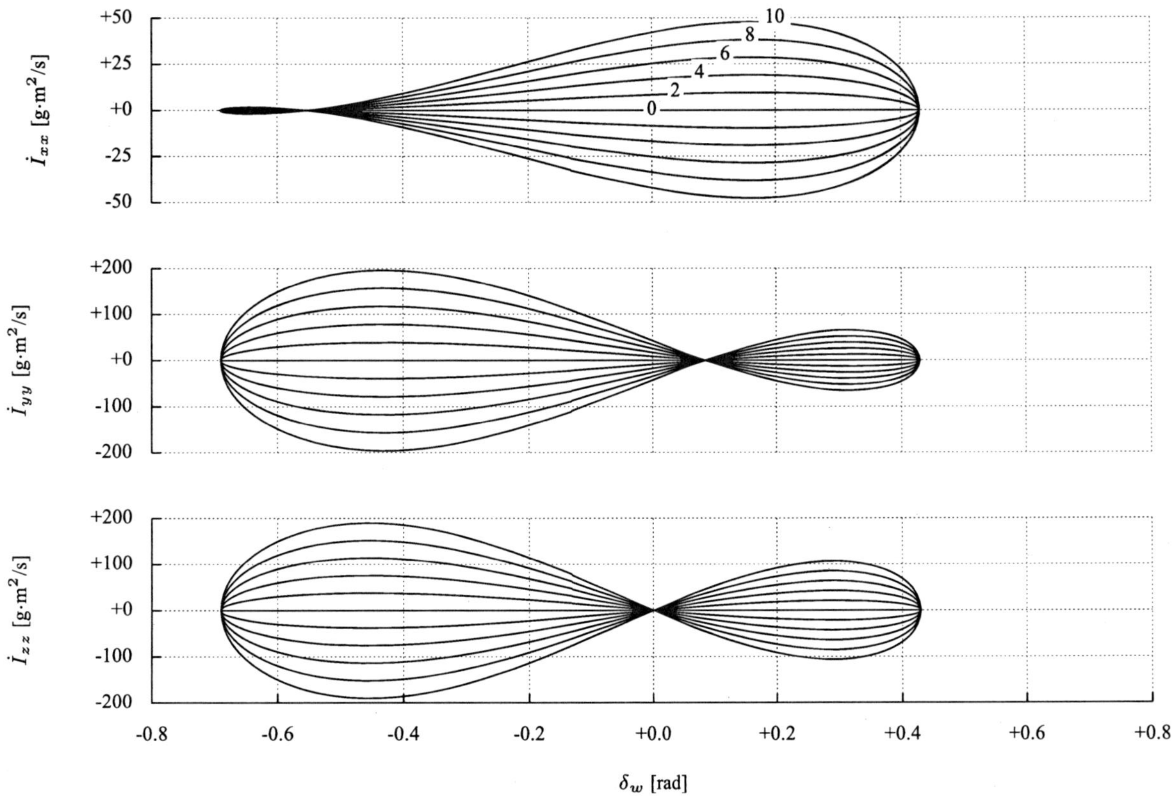

Figure 2.9 **Moment of inertia rates due to flapping between 0 Hz and 10 Hz**

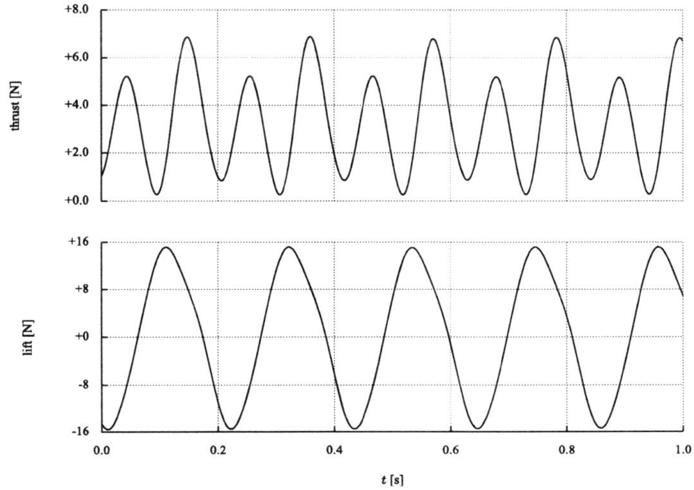

Figure 2.10 **Representative thrust and lift measurements while flapping at 4.7 Hz**

a Reynolds-averaged, Navier-Stokes solver called UMTURNS [57] and a deforming grid algorithm to capture the wing deformations [31]. Good agreement was found at the low flapping frequencies; however at higher frequencies inaccuracies were found in the thrust model. The second analysis was a blade element model using aerodynamics from first principles [19, 20]. It was found that the forward section of the wing creates the lift while the aft section creates the thrust. A quasi-steady aerodynamics model could predict the lift force but not the thrust force.

2.6 Implications for flight dynamics modeling

The flight data showed the presence of fast and large amplitude motions, as well as limit cycle oscillations. Additionally, the harmonic responses measured by the sensor are representative of nonlinear systems or structural vibrations. The large heave accelerations preclude the use of conventional attitude determination algorithms [58] and require model-based state estimators to estimate orientation.

The flapping wings generate a moving center of mass, a variable inertia tensor, significant inertia rates, and a moving reference frame. These difficulties can be modeled in a straight forward manner using multibody system techniques.

The quasi-hover experiment provided a priori information on the thrust and lift waveforms that can be expected in free flight. Modeling results showed that quasi-steady models may be sufficient to model the lift force.

A block diagram for the proposed ornithopter flight dynamics model is shown in Figure 2.11. The pilot control inputs μ are the flapping rate and the tail angles, which impact the aerodynamics. These forces and torques, along with gravitational effects, force the rigid body dynamics, which evolve the generalized position $\mathbf{p}$ and velocity $\mathbf{v}$ states. The rigid body dynamics are nonlinear multibody equations of motion which will be developed in Chapter 3, along with the gravitational effects. The aerodynamic effects will be determined using system identification techniques on wind tunnel and free flight data in Chapter 4. Additional actuator models for this ornithopter have been determined using system identification and are provided in Appendix A.2.

2.7 Chapter summary

This chapter presented an investigation into the flight characteristics of an experimental flapping-wing ornithopter test platform. A flight test was conducted where the ornithopter was fitted with a custom avionics package. Results showed that the ornithopter experiences harmonic responses, large amplitude pitch excursions, fast pitch rates, and large heave accelerations. Geometry modeling using a CAD software showed that as the ornithopter experiences large changes in the vertical position of the center of mass, inertia tensor, and inertia rates as the wings flap. An experiment in a quasi-hover condition showed small deformations of the wing spars, complex flow structures, and the effectiveness

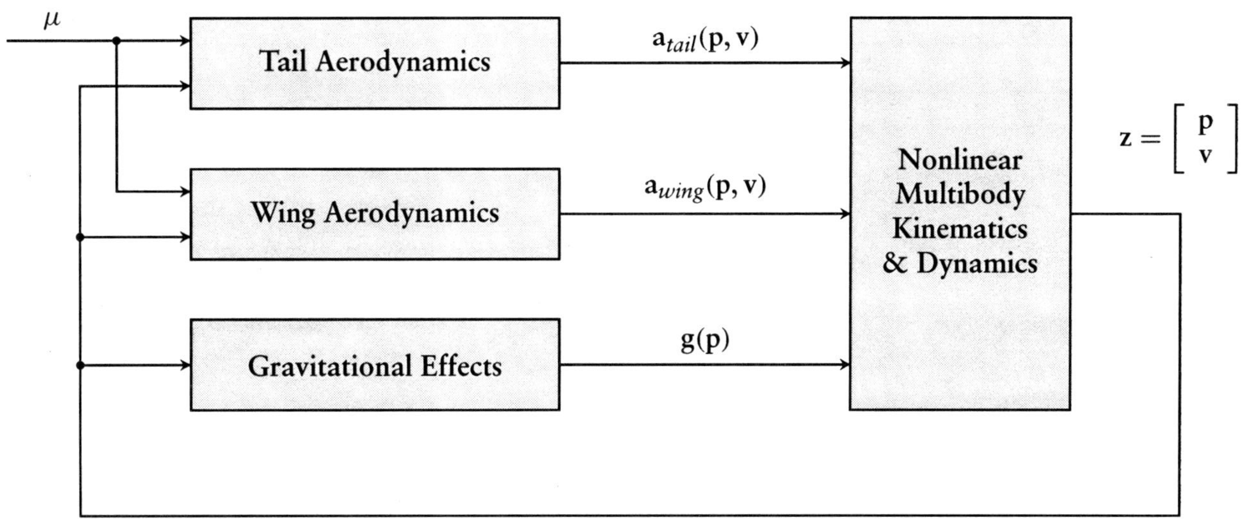

Figure 2.11 Ornithopter open-loop flight dynamics model block diagram

of a quasi-steady blade element model of the aerodynamics. Findings in this chapter suggest a nonlinear multibody model of the ornithopter vehicle dynamics and a quasi-steady model of the aerodynamics.

3

Rigid multibody vehicle dynamics

DOI: 10.1533/9780857094674.37

Abstract: Chapter 2 concluded with a recommendation to model ornithopter flight dynamics using nonlinear multibody techniques. In this chapter, the nonlinear equations of motion are derived for a multibody ornithopter. The model configuration is first presented, where the ornithopter is simplified as a system of three rigid bodies. The kinematic equations are then derived for this system, followed by the dynamic equations. Energy methods were employed, which resulted in a minimum state model, cast into a canonical form and having Lyapunov functions derived from scalar energy functions. At the conclusion of this chapter, only the aerodynamics model remains to be determined.

Key Words: rigid multibody dynamics, Lagrange, equations of motion

3.1 Model configuration

The ornithopter is assumed to be a set of three rigid bodies, with one body allocated for the fuselage and one for each wing, all connected by revolute joints. Although susceptible to bending and torsion, the fuselage is assumed rigid for control modeling purposes. Similarly, the wings are flexible membranes and the aft section deflects to create thrust; however, the wings are assumed to be a rigid body for control modeling, and the aft section contains only 12% of the

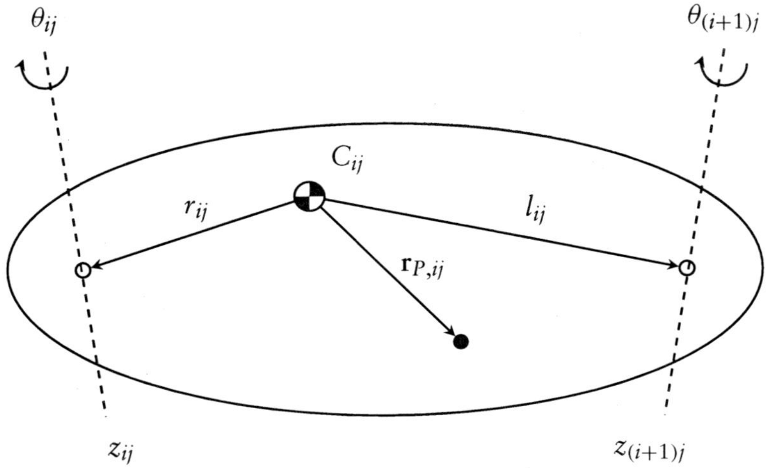

Figure 3.1 **Generic rigid body linkage *i* on chain *j***

wing mass, which is 2.3% of the vehicle mass. The mass and inertia of the tail was absorbed into the fuselage, as the tail comprises only 4.9% of the vehicle mass, is actuated slowly (see Appendix A.2), and adds considerable complexity to the multibody model. Similarly, the four-bar linkage that flaps the wings was not modeled because although it does integrate the flapping kinematics into the equations of motion, it does not contribute a significant source of momentum and adds considerable complexity. Momentum contributions from spinning gears were also neglected.

A generic rigid body, linkage i along chain j, is shown in Figure 3.1. Each rigid body has at its center of mass C_{ij} a mass m_{ij} and inertia tensor $\mathbf{I}_{ij}$. Position vectors l_{ij} and r_{ij} describe the locations of in-board and out-board joint locations, which rotate about vectors z_{ij} and $z_{(i+1)j}$ through angles θ_{ij} and $\theta_{(i+1)j}$. Additionally, an arbitrary point P can be specified by the vector $\mathbf{r}_{P,ij}$.

The ornithopter multibody configuration is shown in Figure 3.2 with mass and inertia properties listed in Table 3.1. The generalized position vector is

$$\mathbf{p} = \begin{bmatrix} \mathbf{r}_{0,I}^{I} \\ \boldsymbol{\eta}_{0,I}^{I} \\ \delta_w \end{bmatrix} \tag{3.1}$$

where the first term is the inertial position of fuselage, the second term is a quaternion describing the orientation of the fuselage, and where the last term

$$\delta_w = \theta_{11} = \pi - \theta_{12} \tag{3.2}$$

is the wing angle. The generalized velocity vector is

$$\mathbf{v} = \begin{bmatrix} \boldsymbol{\nu}_{0,I}^{0} \\ \boldsymbol{\omega}_{0,I}^{0} \\ \dot{\delta}_w \end{bmatrix} \tag{3.3}$$

where the first two terms are the body-fixed translational and rotational velocities of the fuselage, and where the last term is the wing rate. This choice of state variables is the same as for conventional aircraft, except that an additional degree of freedom is provided for the wings.

3.2 Kinematic equations of motion

The kinematic equations describe the evolution of the position state as a function of the velocity state. A generic point P on link i of chain j has the position vector

$$\boldsymbol{r}_{P,I}^{I} = \boldsymbol{r}_{0,I}^{I} + \boldsymbol{R}^{I,0}\boldsymbol{r}_{0j}^{0} + \sum_{k=1}^{i-1} \boldsymbol{R}^{I,kj}(\boldsymbol{r}_{kj}^{kj} - \boldsymbol{l}_{kj}^{kj}) + \boldsymbol{R}^{I,ij}(\boldsymbol{r}_{P,ij}^{ij} - \boldsymbol{l}_{ij}^{ij}) \tag{3.4}$$

expressed in the inertial reference frame. The rotational velocity of a linkage in the ornithopter system can be written in its body frame as

$$\boldsymbol{\omega}_{ij,I}^{ij} = \mathrm{R}^{ij,0}\boldsymbol{\omega}_{0,I}^{0} + \sum_{k=1}^{i} z_{kj}^{kj}\dot{\theta}_{kj}$$

Table 3.1 **Multibody model mass properties**

Index	Mass (kg)	Inertia Tensor $(kg \cdot m^2) \times 10^{-6}$					
ij	m	I_{xx}	I_{yy}	I_{zz}	I_{xy}	I_{xz}	I_{yz}
00	0.3412	+112.57	+3799.3	+3739.4	−51.394	+5.3746	−1.1234
11	0.0414	+915.06	+422.48	+1337.5	−166.07	−0.0001	−0.0000
12	0.0414	+915.06	+422.48	+1337.5	−166.07	−0.0001	−0.0000

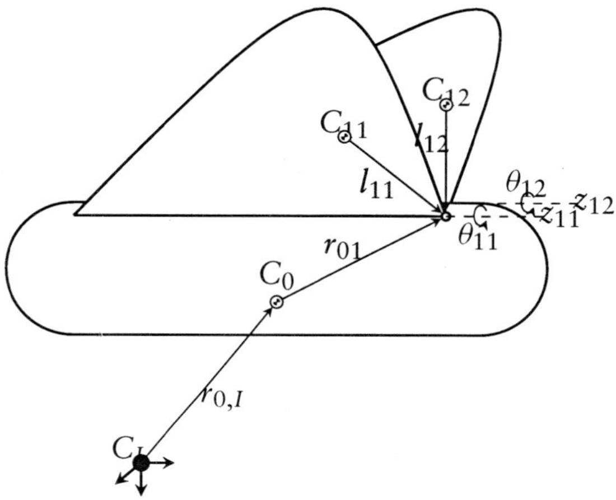

Figure 3.2 **Multibody ornithopter model schematic**

$$= \quad \mathrm{R}^{ij,0}\boldsymbol{\omega}^{0}_{0,I} + \mathrm{Z}^{ij}_{ij}\dot{\boldsymbol{\theta}} \tag{3.5}$$

which forms a convenient way of writing the equations of motion. The translational velocity of a point P on linkage i of chain j is

$$\begin{aligned} \boldsymbol{v}^{0}_{P,I} &= \boldsymbol{v}^{0}_{0,I} + S(\boldsymbol{\omega}^{0}_{0,I})\boldsymbol{r}^{0}_{0j} + \sum_{k=1}^{i-1} S(\boldsymbol{\omega}^{0}_{kj,I})\mathrm{R}^{0,kj}(\boldsymbol{r}^{kj}_{kj} - \boldsymbol{l}^{kj}_{kj}) \\ &\quad + S(\boldsymbol{\omega}^{0}_{ij,I})\mathrm{R}^{0,ij}(\mathbf{r}^{ij}_{p,ij} - \boldsymbol{l}^{ij}_{ij}) \end{aligned} \tag{3.6}$$

expressed in the fuselage frame of reference.

The kinematic differential equation relating the position and velocity states is

$$\dot{p} = \Phi v \tag{3.7}$$

where

$$\boldsymbol{\Phi} = \begin{bmatrix} \mathrm{R}^{I,0} & \\ & \mathrm{J}^{I,0}_{\eta} \\ & \end{bmatrix} \tag{3.8}$$

is a block-diagonal Jacobian matrix with an inverse relation Ψ. The first entry is a rotation matrix, defined using either (2.3) or (2.5), that rotates the translational velocities of the fuselage frame into the inertial coordinate frame. The second term is a rotational Jacobian matrix, defined as either

$$\mathbf{J}_{\eta}^{I,0} = \begin{bmatrix} 1 & \sin\phi\tan\theta & \cos\phi\tan\theta \\ 0 & \cos\phi & -\sin\phi \\ 0 & \sin\phi\cos\theta & \cos\phi\cos\theta \end{bmatrix} \quad (3.9)$$

or

$$\mathbf{J}_{\eta}^{I,0} = \frac{1}{2}\begin{bmatrix} \delta + S(\epsilon) \\ -\epsilon^T \end{bmatrix} \quad (3.10)$$

and relates body-fixed velocities to attitude rates. The last entry is an identity matrix, reflecting the kinematically flat relationship of the joint angles and their rates.

3.3 Dynamic equations of motion

The dynamic equations describe the evolution of the velocity as a function of the states and controls. These equations are derived here using energy methods, which produce first-order state space equations with a minimal number of state variables. As aircraft dynamics are not typically derived using energy methods, Appendix B.3 is included to provide detail and familiarity.

The kinetic and potential energy, which may serve as future Lyapunov control functions [59], are

$$T(\mathbf{p},\mathbf{v}) = \frac{1}{2}\sum_{i=0}^{n_i}\sum_{j=0}^{n_j} m_{ij}(\boldsymbol{\nu}_{ij,I}^{0})^T(\boldsymbol{\nu}_{ij,I}^{0}) + (\boldsymbol{\omega}_{ij,I}^{ij})^T\mathbf{I}_{ij}^{ij}(\boldsymbol{\omega}_{ij,I}^{ij}) \quad (3.11)$$

$$U(\mathbf{p}) = -\sum_{i=0}^{n_i}\sum_{j=0}^{n_j} m_{ij}(\mathbf{r}_{ij,I}^{I})^T\mathbf{g}^I \quad (3.12)$$

where n_j are the total number of kinematic chains, n_i are the number of linkages on chain j, and g is the local acceleration due to gravity.

The Boltzmann-Hamel equations [60, 61], a generalization of the Lagrange equations which account for different reference frames,

$$\frac{d}{dt}\left[\frac{\partial T}{\partial \boldsymbol{v}}\right]^T + \left(\sum_{k=1}^{n_v} \frac{\partial T}{\partial v_k}\boldsymbol{\Gamma}_k\right)\mathrm{v} - \boldsymbol{\Phi}^T\left[\frac{\partial T}{\partial \boldsymbol{p}}\right]^T = \boldsymbol{\tau} \tag{3.13}$$

may then be applied, where potential energy sources are neglected and the Hamel coefficient matrices are defined as

$$\boldsymbol{\Gamma}_k = \boldsymbol{\Phi}^T \boldsymbol{\Lambda}_k \boldsymbol{\Phi} \tag{3.14}$$

with individual matrix elements

$$\{\boldsymbol{\Lambda}_k\}_{ij} = \frac{\partial \Phi_{ki}}{\partial p_j} - \frac{\partial \Psi_{kj}}{\partial p_i}. \tag{3.15}$$

For a given multibody system and choice of reference frames, the relationship between the velocities and quasi-velocities of the system is fixed, and so the Hamel coefficient matrices (3.14) are constant, skew symmetric matrices.

The equations (3.13) can be manipulated into a canonical form used in the nonlinear control of Euler-Lagrange systems [62, 63, 64, 59]. The kinetic energy of the system can be written

$$T(\mathbf{p}, \mathbf{v}) = \frac{1}{2}\mathbf{v}^T \mathbf{M}(\mathbf{p})\mathbf{v} \tag{3.16}$$

which is a quadratic form in the velocity. Inserting (3.16) into (3.13), differentiating, and regrouping the terms, yields the reformulation

$$\mathrm{M}\dot{\mathrm{v}} + \left(\dot{\mathrm{M}} + \sum_{k=1}^{n_v} \frac{\partial T}{\partial v_k}\boldsymbol{\Gamma}_k - \frac{1}{2}\boldsymbol{\Phi}^T(\mathbb{I} \otimes \mathbf{v}^T)\left[\frac{\partial \mathbf{M}}{\partial \mathbf{p}}\right]^T\right)\mathrm{v} = \boldsymbol{\tau} \tag{3.17}$$

where $\otimes$ is a Kronecker product operator [65].

Grouping terms in (3.17) and adding forces imparted on the vehicle by the external environment yields the canonical form

$$M(p)\dot{v} + C(p, v)v + g(p) + a(p, v) = \tau \qquad (3.18)$$

where $\mathbf{M(p)}$ is a generalized mass matrix, $\mathbf{C(p, v)}$ contains nonlinear coupling forces arising from centripetal and Coriolis accelerations, $\mathbf{g(p)}$ describes gravitational effects, and $\mathbf{a(p, v)}$ describes aerodynamic effects. With this choice of state variables, the forcing vector τ corresponds to forces on the fuselage center of mass, torques on the fuselage center of mass, and torques on the wings. In the following sections these matrices and vectors are derived and discussed, with the exception of the aerodynamic terms, which are presented in Chapter 4. Properties of the canonical form are further discussed in Kelly et al. [63] and Lewis et al. [62].

3.3.1 Inertial effects

The mass matrix $\mathbf{M(p)}$ describes the generalized mass and inertia of the system as a function of the vehicle pose. Contributions to this matrix can be determined for each rigid body by using algebraic manipulations to rearrange the kinetic coenergy (3.11) into the quadratic form (3.16), for which the mass matrix is then obvious.

The mass matrix for the fuselage body is found by rearranging the energy and has the same form as that found in conventional aircraft equations of motion [1, 2]. The mass matrix is

$$\mathbf{M(p)} = \begin{bmatrix} ||m_0 & & \\ & \mathbf{I}_0^0 & \\ & & \end{bmatrix} \qquad (3.19)$$

which is a diagonal matrix with the mass and inertia tensor along the block diagonal. The mass matrix for the first linkage ($i = 1$, e.g. the wings) is given in (3.20), and should a second linkage ($i = 2$, e.g. the four-bar mechanism, tail surface, etc.) need to be modeled, the mass matrix is given in (3.21). Each mass matrix has along its block diagonal the scalar mass, the inertia relative to the fuselage center of mass, and the inertia of the linkage. Off-diagonal terms represent inertial couplings within the system. The total mass matrix of the system is a summation of the mass matrices of each rigid body.

3.3.2 Nonlinear coupling effects

The nonlinear coupling matrix is a non-unique, square matrix that contains forces arising from Coriolis and centripetal accelerations, and is most easily derived from the mass matrix. It is apparent that one definition of the coupling matrix is

$$\mathbf{C}(\mathbf{p},\mathbf{v}) = \dot{\mathbf{M}} + \sum_{k=1}^{n_v} \frac{\partial T}{\partial v_k}\mathbf{\Gamma}_k - \frac{1}{2}\mathbf{\Phi}^T(\mathbb{I}\otimes\mathbf{v}^T)\left[\frac{\partial \mathbf{M}}{\partial \mathbf{p}}\right]^T \quad (3.22)$$

from comparing (3.17) and (3.18). A property exploited in passivity-based control is that there exists a representation of the coupling matrix that satisfies

$$\dot{\mathbf{M}}(p,v) - 2\mathbf{C}(\mathbf{p},\mathbf{v}) \quad (3.23)$$

is a skew-symmetric matrix [64, 66]. The representation (3.22) does not satisfy this relation; rather, by extension of the discussion by Lewis et al. [62], the representation is

$$\begin{aligned}\mathbf{C}(\mathbf{p},\mathbf{v}) &= \tfrac{1}{2}\dot{\mathbf{M}} + \left(\textstyle\sum_{k=1}^{n_v} \frac{\partial T}{\partial v_k}\mathbf{\Gamma}_k\right) \\ &+\tfrac{1}{2}\left[\tfrac{\partial \mathbf{M}}{\partial \mathbf{p}}\right](\mathbf{v}\otimes\mathbb{I})\mathbf{\Phi} - \tfrac{1}{2}\mathbf{\Phi}^T(\mathbb{I}\otimes\mathbf{v}^T)\left[\tfrac{\partial \mathbf{M}}{\partial \mathbf{p}}\right]^T \end{aligned} \quad (3.24)$$

so that

$$\dot{\mathbf{M}}-2\mathbf{C} = -2\left(\sum_{k=1}^{n_v} \frac{\partial T}{\partial v_k}\mathbf{\Gamma}_k\right) - \left[\frac{\partial \mathbf{M}}{\partial \mathbf{p}}\right](\mathbf{v}\otimes\mathbb{I})\mathbf{\Phi} + \mathbf{\Phi}^T(\mathbb{I}\otimes\mathbf{v}^T)\left[\frac{\partial \mathbf{M}}{\partial \mathbf{p}}\right]^T \quad (3.25)$$

which satisfies the property (3.23) because the first term is a summation of skew Hamel coefficient matrices and the last two terms represent the skew decomposition of a matrix.

Evaluating the coupling matrix (3.24) for a given ornithopter state is not a trivial task. The mass matrix derivatives, both respect to time and position, were derived. Some

$$
\mathbf{M}(\mathbf{p}) = \begin{pmatrix}
||m_{1j}\mathbb{I} & m_{1j}S(\mathbf{R}^{0,1j}l_{1j}^{1j}) - m_{1j}S(r_{0j}^{0}) & m_{1j}S(\mathbf{R}^{0,1j}l_{1j}^{1j})\mathbf{Z}_{1j}^{0} \\
\cdots & \begin{array}{c} m_{1j}\left[(r_{0j}^{0})^{T}(r_{0j}^{0}) - 2(l_{1j}^{1j})^{T}\mathbf{R}^{1j,0}r_{0j}^{0} + (l_{1j}^{1j})^{T}(l_{1j}^{1j})\right]\mathbb{I} \\ -m_{1j}(r_{0j}^{0})(r_{0j}^{0})^{T} + m_{1j}\mathbf{R}^{0,1j}l_{1j}^{1j}(r_{0j}^{0})^{T} \\ +m_{1j}r_{0j}^{0}(l_{1j}^{1j})^{T}\mathbf{R}^{1j,0} - m_{1j}\mathbf{R}^{0,1j}(l_{1j}^{1j})(l_{1j}^{1j})^{T}\mathbf{R}^{1j,0} \\ +\mathbf{R}^{0,1j}\mathbf{I}_{1j}^{1j}\mathbf{R}^{1j,0} \end{array} & \begin{array}{c} m_{1j}\left[(l_{1j}^{1j})^{T}(l_{1j}^{1j}) - (l_{1j}^{1j})^{T}\mathbf{R}^{1j,0}r_{0j}^{0}\right]\mathbf{Z}_{1j}^{0} \\ +m_{1j}\mathbf{R}^{0,1j}l_{1j}^{1j}(r_{0j}^{0})^{T}\mathbf{Z}_{1j}^{0} \\ -m_{1j}\mathbf{R}^{0,1j}l_{1j}^{1j}(l_{1j}^{1j})^{T}\mathbf{R}^{1j,0}\mathbf{Z}_{1j}^{0} \\ +\mathbf{R}^{0,1j}\mathbf{I}_{1j}^{1j}\mathbf{Z}_{1j}^{0} \end{array} \\
\cdots & \cdots & \begin{array}{c} m_{1j}(\mathbf{Z}_{1j}^{0})^{T}\left[(l_{1j}^{1j})^{T}(l_{1j}^{1j})\right]\mathbf{Z}_{1j}^{0} \\ -m_{1j}(\mathbf{Z}_{1j}^{0})^{T}\mathbf{R}^{0,1j}l_{1j}^{1j}(l_{1j}^{1j})^{T}\mathbf{R}^{1j,0}\mathbf{Z}_{1j}^{0} \\ +(\mathbf{Z}_{1j}^{1j})^{T}\mathbf{I}_{1j}^{1j}\mathbf{Z}_{1j}^{1j} \end{array}
\end{pmatrix} \tag{3.20}
$$

$$
\mathbf{M}(\mathbf{p}) = \begin{pmatrix}
||m_{2j}\mathbb{I} & \begin{array}{c} -m_{2j}S(r^0_{0j}) - m_{2j}S(\mathbf{R}^{0,1j}(r^{1j}_{1j} - l^{1j}_{1j})) \\ +m_{2j}S(\mathbf{R}^{0,2j}l^{2j}_{2j}) \end{array} & \begin{array}{c} -m_{2j}S(\mathbf{R}^{0,1j}(r^{1j}_{1j} - l^{1j}_{1j}))\mathbf{Z}^0_{1j} \\ +m_{2j}S(\mathbf{R}^{0,2j}l^{2j}_{2j})\mathbf{Z}^0_{2j} \end{array} \\
\cdots & \begin{array}{c} m_{2j}\left[(r^0_{0j})^T(r^0_{0j})\right]\mathbb{I} - m_{2j}(r^0_{0j})(r^0_{0j})^T \\ +m_{2j}\left[(r^0_{0j})^T\mathbf{R}^{0,1j}(r^{1j}_{1j} - l^{1j}_{1j})\right]\mathbb{I} - m_{2j}r^0_{0j}(r^{1j}_{1j} - l^{1j}_{1j})^T\mathbf{R}^{1j,0} \\ +\mathrm{m}_{2j}\mathrm{r}^0_{0j}(\mathrm{l}^{2j}_{2j})^T\mathbf{R}^{2j,0} - \mathrm{m}_{2j}\left[(\mathrm{r}^0_{0j})^T\mathbf{R}^{0,2j}\mathrm{l}^{2j}_{2j}\right]\mathbb{I} \\ +m_{2j}\left[(r^{1j}_{1j} - l^{1j}_{1j})^T\mathbf{R}^{1j,0}r^0_{0j}\right]\mathbb{I} - m_{2j}\mathbf{R}^{0,1j}(r^{1j}_{1j} - l^{1j}_{1j})(r^0_{0j})^T \\ +m_{2j}(r^{1j}_{1j} - l^{1j}_{1j})^T(r^{1j}_{1j} - l^{1j}_{1j})\mathbb{I} \\ -m_{2j}\mathbf{R}^{0,1j}(r^{1j}_{1j} - l^{1j}_{1j})(r^{1j}_{1j} - l^{1j}_{1j})^T\mathbf{R}^{1j,0} \\ +m_{2j}\mathbf{R}^{0,1j}(r^{1j}_{1j} - l^{1j}_{1j})(l^{2j}_{2j})^T\mathbf{R}^{2j,0} \\ -m_{2j}\left[(r^{1j}_{1j} - l^{1j}_{1j})^T\mathbf{R}^{1j,2j}l^{2j}_{2j}\right]\mathbb{I} + m_{2j}\mathbf{R}^{0,2j}(l^{2j}_{2j})(r^0_{0j})^T \\ -m_{2j}\left[(l^{2j}_{2j})^T\mathbf{R}^{2j,0}r^0_{0j}\right]\mathbb{I} + m_{2j}\mathbf{R}^{0,2j}l^{2j}_{2j}(r^{1j}_{1j} - l^{1j}_{1j})^T\mathbf{R}^{1j,0} \\ -m_{2j}\left[(l^{2j}_{2j})^T\mathbf{R}^{2j,1j}(r^{1j}_{1j} - l^{1j}_{1j})\right]\mathbb{I} + m_{2j}\left[(l^{2j}_{2j})^T l^{2j}_{2j}\right]\mathbb{I} \\ -m_{2j}\mathbf{R}^{0,2j}(l^{2j}_{2j})(l^{2j}_{2j})^T\mathbf{R}^{2j,0} + \mathbf{R}^{0,2j}\mathbf{I}^{2j}_{2j}\mathbf{R}^{2j,0} \end{array} & \begin{array}{c} m_{2j}\left[(r^{1j}_{1j} - l^{1j}_{1j})^T\mathbf{R}^{1j,0}r^0_{r0j}\right]\mathbf{Z}^0_{1j} \\ -m_{2j}\mathbf{R}^{0,1j}(r^{1j}_{1j} - l^{1j}_{1j})(r^0_{0j})^T\mathbf{Z}^0_{1j}+ \\ m_{2j}\left[(r^{1j}_{1j} - l^{1j}_{1j})^T(r^{1j}_{1j} - l^{1j}_{1j})\right]\mathbf{Z}^0_{1j} \\ +m_{2j}\mathbf{R}^{0,1j}(r^{1j}_{1j} - l^{1j}_{1j})(r^{1j}_{1j} - l^{1j}_{1j})^T\mathbf{R}^{1j,0}\mathbf{Z}^0_{1j} \\ +m_{2j}\mathbf{R}^{0,1j}(r^{1j}_{1j} - l^{1j}_{1j})(l^{2j}_{2j})^T\mathbf{R}^{2j,0}\mathbf{Z}^0_{1j} \\ -m_{2j}\left[(r^{1j}_{1j} - l^{1j}_{1j})^T\mathbf{R}^{1j,2j}l^{2j}_{2j}\right]\mathbf{Z}^0_{1j}, \\ +m_{2j}\mathbf{R}^{0,2j}(l^{2j}_{2j})(r^0_{0j})^T\mathbf{Z}^0_{2j} \\ -m_{2j}\left[(l^{2j}_{2j})^T\mathbf{R}^{2j,0}r^0_{0j}\right]\mathbf{Z}^0_{2j} \\ +m_{2j}\mathbf{R}^{0,2j}l^{2j}_{2j}(r^{1j}_{1j} - l^{1j}_{1j})^T\mathbf{R}^{1j,0}\mathbf{Z}^0_{2j} \\ -m_{2j}\left[(l^{2j}_{2j})^T\mathbf{R}^{2j,1j}(r^{1j}_{1j} - l^{1j}_{1j})\right]\mathbf{Z}^0_{2j} \\ +m_{2j}\left[(l^{2j}_{2j})^T(l^{2j}_{2j})\right]\mathbf{Z}^0_{2j} \\ -m_{2j}\mathbf{R}^{0,2j}(l^{2j}_{2j})(l^{2j}_{2j})^T\mathbf{R}^{2j,0}\mathbf{Z}^0_{2j}\mathbf{R}^{0,2j}\mathbf{I}^{2j}_{2j}\mathbf{Z}^{2j}_{2j} \end{array} \\
\cdots & \cdots & \begin{array}{c} m_{2j}(\mathbf{Z}^0_{1j})^T(r^{1j}_{1j} - l^{1j}_{1j})^T(r^{1j}_{1j} - l^{1j}_{1j})\mathbf{Z}^0_{1j} \\ -m_{2j}(\mathbf{Z}^0_{1j})^T\mathbf{R}^{0,1j}(r^{1j}_{1j} - l^{1j}_{1j})(r^{1j}_{1j} - l^{1j}_{1j})^T\mathbf{R}^{1j,0}\mathbf{Z}^0_{1j} \\ +m_{2j}(\mathbf{Z}^0_{2j})^T\mathbf{R}^{0,1j}(r^{1j}_{1j} - l^{1j}_{1j})(l^{2j}_{2j})^T\mathbf{R}^{2j,0}\mathbf{Z}^0_{1j} \\ -m_{2j}(\mathbf{Z}^0_{2j})^T\left[(r^{1j}_{1j} - l^{1j}_{1j})^T\mathbf{R}^{1j,2j}l^{2j}_{2j}\right]\mathbf{Z}^0_{1j} \\ +m_{2j}(\mathbf{Z}^0_{1j})^T\mathbf{R}^{0,2j}l^{2j}_{2j}(r^{1j}_{1j} - l^{1j}_{1j})^T\mathbf{R}^{1j,0}\mathbf{Z}^0_{2j} \\ -m_{2j}(\mathbf{Z}^0_{1j})^T\left[(l^{2j}_{2j})^T\mathbf{R}^{2j,1j}(r^{1j}_{1j} - l^{1j}_{1j})\right]\mathbf{Z}^0_{2j} \\ +m_{2j}(\mathbf{Z}^0_{2j})^T\left[(l^{2j}_{2j})^T(l^{2j}_{2j})\right]\mathbf{Z}^0_{2j} \\ -m_{2j}(\mathbf{Z}^0_{2j})^T\mathbf{R}^{0,2j}(l^{2j}_{2j})(l^{2j}_{2j})^T\mathbf{R}^{2j,0}\mathbf{Z}^0_{2j} \\ +(\mathbf{Z}^{2j}_{2j})^T\mathbf{I}^{2j}_{2j}\mathbf{Z}^{2j}_{2j} \end{array}
\end{pmatrix} \tag{3.21}
$$

simplifications of the equations arise, as the Kronecker products have many null entries and the mass matrix is only a function of the joint position states.

3.3.3 Gravitational effects

Gravitation effects depend on the orientation of the ornithopter, and are derived from the potential energy as

$$\mathbf{g}(\mathbf{p}) = \mathbf{\Phi}^T \left[\frac{\partial U}{\partial \mathbf{p}} \right]^T \tag{3.26}$$

however, this formulation leads to complicated expressions involving tensors. It is simpler in this case to include gravitational effects as external forces applied at the centers of mass

$$\mathbf{g}(\mathbf{p}) = \sum_{i=0}^{n_i} \sum_{j=0}^{n_j} \mathbf{J}_{f,ij} \begin{bmatrix} 0 \\ 0 \\ m_{ij} g^I \end{bmatrix} \tag{3.27}$$

where the force Jacobian matrices are

$$\mathbf{J}_{f,0} = \begin{bmatrix} \mathbf{R}^{0,I} \\ 0 \\ 0 \end{bmatrix} \tag{3.28}$$

for the fuselage $i = 0$,

$$\mathbf{J}_{f,1j} = \begin{bmatrix} \mathbf{R}^{0,I} \\ S(r^0_{0j} - \mathbf{R}^{0,1j} l^{1j}_{1j}) \\ (\mathbf{Z}^{1j}_{1j})^T S(-l^{1j}_{1j}) \mathbf{R}^{1j,I} \end{bmatrix} \tag{3.29}$$

for the first linkage $i = 1$ on a chain j, and

$$\mathbf{J}_{f,2j} = \begin{bmatrix} \mathbf{R}^{0,I} \\ S(r^0_{0j} + \mathbf{R}^{0,1j}(r^{1j}_{1j} - l^{1j}_{1j}) - \mathbf{R}^{0,2j} l^{2j}_{2j}) \mathbf{R}^{0,I} \\ (\mathbf{Z}_{1j})^T S((r^{1j}_{1j} - l^{1j}_{1j}) - \mathbf{R}^{1j,2j} l^{2j}_{2j}) \mathbf{R}^{1j,I} + (\mathbf{Z}^{2j}_{2j})^T S(-l^{2j}_{2j}) \mathbf{R}^{2j,I} \end{bmatrix} \tag{3.30}$$

for the second linkage $i = 2$ on a chain j.

3.4 Chapter summary

In this chapter, the nonlinear multibody equations of motion for an ornithopter were derived. A rigid body is allocated for the fuselage and each wing. Kinematic and dynamic equations of motion are presented. The Boltzmann-Hamel equations were used, which result in Lyapunov control functions and a minimal state space model that was cast into a canonical form. Gravitational effects were also derived. At this point, the only unmodeled forces on the ornithopter are due to aerodynamics, which will be presented in Chapter 4.

4

System identification of aerodynamic models

DOI: 10.1533/9780857094674.51

Abstract: In Chapter 2 it was determined that a nonlinear multibody model was needed to describe the flight dynamics of the ornithopter. In Chapter 3 the rigid body equations of motion, as well as the gravitational effects, were derived from first principles. The aerodynamics for an arbitrary flapping wing vehicle are difficult to model analytically, and are determined in this work using system identification methods with experimental data. This chapter begins with a presentation of the system identification methods employed herein. Afterwards, tail aerodynamic models are identified from wind tunnel data, and wing aerodynamic models are identified from flight test data. This chapter concludes the nonlinear multibody model of the ornithopter flight dynamics.

Key Words: system identification, aerodynamics, nonlinear

4.1 System identification method

System identification is the process of determining, using observations of inputs and outputs, a mathematical model that behaves similarly to the physical system under specified conditions. Applied to aircraft, system identification can be used to create flight simulators, validate numerical codes, design control laws, assess handling qualities, perform model reduction, and gain physical insight into the system. In this work, system identification was employed to develop aerodynamic models for the ornithopter that could not be determined from first principles modeling. Several texts provide a thorough description of system identification techniques [67, 37, 68].

The system identification process begins with an experiment where inputs and outputs are recorded while the system is sufficiently excited. After any necessary calibration, signal processing, data compatibility and data collinearity analyses, the next step is model structure determination, which is the process of finding an adequate functional representation relating inputs to outputs. While there are many techniques for this step, step-wise regression is most heavily relied on in this work [69, 37]. In this method, a pool of candidate regressors are selected and are sequentially included in or excluded from the model per engineering intuition and several statistical metrics. The process is iterative and interactive, and can be applied to linear or nonlinear problems, in the time or frequency domains, using any method of parameter estimation.

Once the form of the model is known, the next task is to estimate the unknown parameters in the model. For the system

$$\dot{\mathbf{z}} = \mathbf{f}(t, \mathbf{z}, \boldsymbol{\mu}) \tag{4.1}$$

with states $\mathbf{z}$, inputs μ, and measurements $\mathbf{y}_m$, the maximum likelihood cost function

$$J(\phi) = \frac{1}{2}\sum_{i=1}^{n} \mathbf{e}^T \mathbf{W}^{-1} \mathbf{e} \tag{4.2}$$

where ϕ is the vector of unknown parameters, $\mathbf{e}$ is the residual vector between the measurements and the outputs, and $\mathbf{W}$ is the inverse noise covariance matrix. Two simplifications of the maximum likelihood estimator are the equation-error and output-error methods [70, 37]. In equation-error, process noise is assumed and the measurements are the state derivatives. This results in a linear estimation problem with an analytical solution. In output-error, measurement noise is assumed and the measurements are the states. This results in a nonlinear, iterative solution that requires initial guesses, but is considered more accurate than equation-error. Once parameter estimates are obtained, a new data set is used to test the predictive capability of the identified model.

All signal processing and system identification algorithms are implemented in this work using a MATLAB toolbox called System IDentification Programs for AirCraft (SIDPAC), developed at NASA Langley Research Center and documented in Reference [37].

4.2 Tail aerodynamics

This section describes the system identification of the ornithopter tail aerodynamic model using wind tunnel data. The tail aerodynamics are assumed to be steady due to limited pilot bandwidth [71] and actuator bandwidths (provided in Appendix A.2), and independent of the wing aerodynamics. These simplifying assumptions are common in the literature due to the complex nature of the interactions [28, 25, 21].

4.2.1 Experimental setup

The ornithopter was tested in a free-jet wind tunnel, having a 0.56 m by 0.56 m test section. The wings were removed and the ornithopter was fitted on a custom test stand in the center of the test section, as shown in Figure 4.1. Free stream velocity measurements were obtained using a U-tube manometer and a barometer. The test stand included a load cell that produced six-axis force and torque data. Tail positions were measured using the internal potentiometers in the servo motors. The tail orientation was controlled using potentiometers and PIC18F452 microprocessors, as well as a signal generator.

Four tests were selected for system identification of aerodynamic force and moment coefficients. Each test occurred at a different free stream velocity, and a potentiometer was used to step the angle of attack on the wings while the sideslip

Figure 4.1 Wind tunnel test experimental setup for tail aerodynamics modeling

angle was cycled at 0.06 Hz using a signal generator. Aerodynamic quantities for these tests are concatenated and shown in Figure 4.2. Flow velocities ranged between 4.50 m/s and 6.33 m/s; the lower bound was the slowest velocity with an acceptable signal to noise ratio, and the upper bound began to exhibit structural deformations and increases in actuator current consumption. The angle of attack range, -0.88 to -0.35 rad, and angle of sideslip range, ±0.57 rad, comprise the full travel of the tail orientation. A frequency domain analysis showed that no significant unsteady effects were present in the data due to the cycling of the sideslip angle, and the associated reduced frequency ranges between 0.006 and 0.009, which is a regime where unsteady effects are not significant [18].

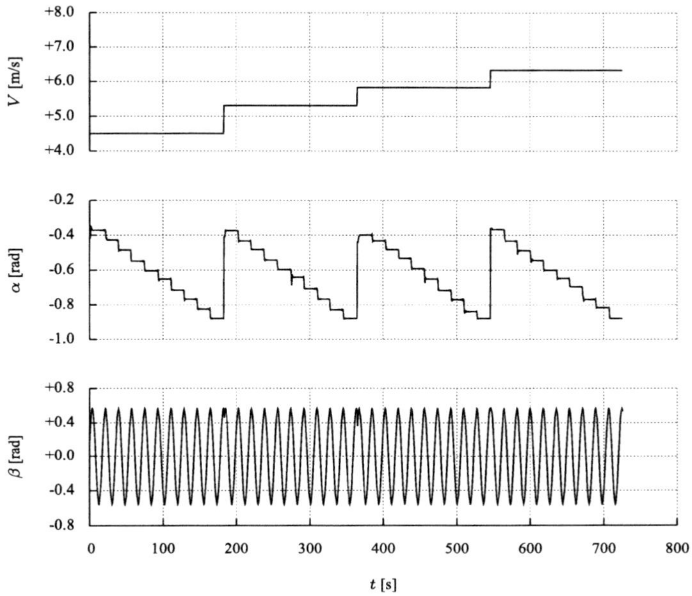

Figure 4.2 **Flow states from four concatenated wind tunnel tests used for modeling the tail aerodynamics**

4.2.2 Results

Data were calibrated, filtered to 1 Hz using a low-pass smoother, and then concatenated together. Aerodynamic force and moment coefficients were computed as

$$\begin{aligned} C_x &= X/QS_t \\ C_y &= Y/QS_t \\ C_z &= Z/QS_t \\ C_l &= L/QS_t b_t \\ C_m &= M/QS_t c_t \\ C_n &= N/QS_t b_t \end{aligned} \tag{4.3}$$

where

$$Q = \frac{1}{2}\rho V^2 \tag{4.4}$$

is the dynamic pressure, X, Y, and Z are the body axis forces, L, M, and N are the body axis moments, and tail geometry parameters are listed in Table 2.1.

Model structure determination was performed using stepwise regression in the time domain using equation-error. The model structure

$$\begin{aligned} C_x &= C_{x_0} + C_{x_{\alpha^2}}\alpha^2 \\ C_y &= C_{y_{\alpha\beta}}\alpha\beta \\ C_z &= C_{z_0} + C_{z_\alpha}\alpha + C_{z_\beta}\beta \\ C_l &= C_{l_\beta}\beta + C_{l_{\alpha\beta}}\alpha\beta \\ C_m &= C_{m_0} + C_{m_\alpha}\alpha + C_{m_\beta}\beta + C_{m_{\beta^2}}\beta^2 + C_{m_{\alpha V}}\alpha V \\ C_n &= C_{n_\beta}\beta \end{aligned} \tag{4.5}$$

was found to fit the data well. The model is nonlinear and expresses the aerodynamic force and moment coefficients in terms of the measured aerodynamic flow variables. The lateral models are zero mean while the longitudinal models contain biases to account for profile drag, as well as lift and pitching moment at zero angle of attack. The model for drag is the traditional drag polar. The yawing moment is dependent only on the sideslip angle. The remaining models contain a coupling between the angle of attack and sideslip angle, due to the serial configuration of the tail surface. The measured yawing moment is much larger than the rolling moment, which indicates the lateral control is primarily in yaw.

Time domain equation-error was used to estimate model parameters, as the model equations precluded an output-error analysis, and testing at steady free stream velocities prohibited an accurate frequency domain analysis. Measured and modeled coefficients are shown in Figure 4.3, and estimated parameters with standard errors are given in Table 4.1. The C_z model had a lower fit value due to the residual seen in the third and fourth data sets, and the C_{z_α} and $C_{m_{\beta^2}}$ coefficients had slightly larger errors, but otherwise the models produced high fit values and parameter errors below 10%, indicating good model fits.

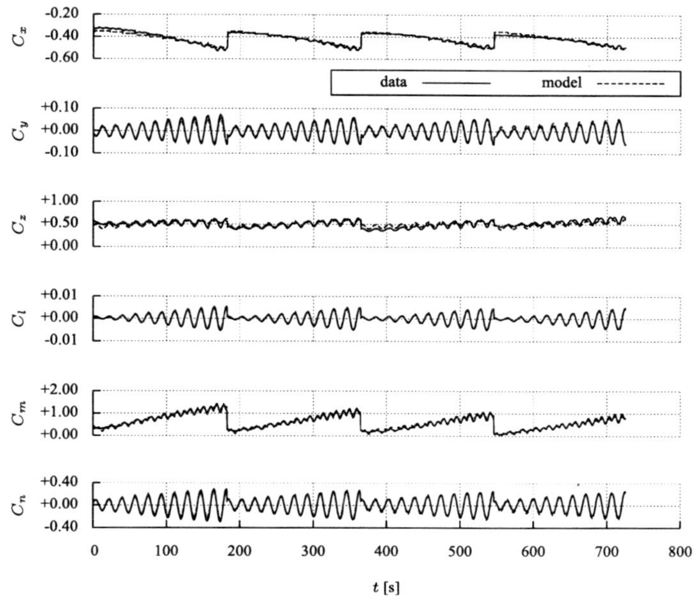

Figure 4.3 **Tail aerodynamic model fits**

A similar test, not used for identification and having a free stream velocity of 4.86 m/s, was used to validate the identified model. Measured and predicted force and moment coefficients are shown in Figure 4.4. The rolling and pitching moment coefficients have small residuals at low angles of attack, but overall the models predict well and all have coefficients of determination above 0.81, indicating a good model.

4.3 Wing aerodynamics

This section describes the system identification of the ornithopter wing aerodynamics model from flight data. The experimental setup is first presented, followed by a discussion of the data reduction and kinematic flight results, and lastly a presentation of the system identification results.

Table 4.1 Tail aerodynamic parameters and standard errors

Parameter	Estimate	Fit
ϕ	$\hat{\phi} \pm \sigma(\hat{\phi})$	R^2
C_{x_0}	-0.3181 ± 0.0058	0.88
$C_{x_{\alpha^2}}$	-0.2310 ± 0.0123	
$C_{y_{\alpha\beta}}$	$+0.1153 \pm 0.0035$	0.88
C_{z_0}	$+0.3346 \pm 0.0243$	
C_{z_α}	-0.2729 ± 0.0364	0.59
C_{z_β}	$+0.0884 \pm 0.0066$	
C_{l_β}	-0.0054 ± 0.0005	0.93
$C_{l_{\alpha\beta}}$	-0.0161 ± 0.0007	
C_{m_0}	-0.3486 ± 0.0272	
C_{m_α}	-3.3182 ± 0.1131	
C_{m_β}	$+0.0975 \pm 0.0067$	0.96
$C_{m_{\beta^2}}$	-0.4184 ± 0.0847	
$C_{m_{\alpha V}}$	$+0.3053 \pm 0.0195$	
$C_{n_{\alpha\beta}}$	-0.5094 ± 0.0141	0.97

4.3.1 Experimental setup

It is difficult to obtain flight data for system identification of an ornithopter. Avionics add weight and inertia to the ornithopter, and must sample faster than with conventional aircraft to measure the effects of flapping. Additionally, the large heave accelerations prohibit an attitude estimation using conventional methods with magnetometers and accelerometers, and air data probes incur inaccuracies due to the fast pitching and heaving motions. Flexibility in the aircraft also adds errors to the responses at harmonics of the flapping frequency. To mitigate these problems, a visual tracking system was used to measure the spatial location of markers placed on the ornithopter. At the cost of restricting the capture volume for data collection, this is a non-invasive and highly accurate method that additionally averages out effects due to flexibility.

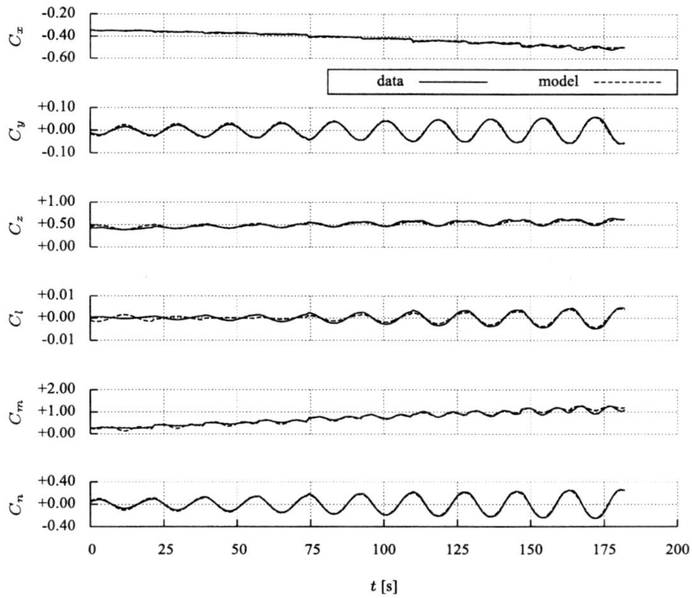

Figure 4.4 **Tail aerodynamics model prediction case**

Flight tests consisted of flying the ornithopter down a corridor, as shown in Figure 4.5, while collecting data. This method was sufficient for modeling straight and level flight while minimizing atmospheric disturbances and facilitating the use of the visual tracking system. Eight cameras were placed at the corners of a capture volume 10 m long by 6 m wide and 5 m tall. These cameras recorded at 500 Hz the spatial position of retro-reflective markers placed on the ornithopter fuselage, wings, and tail, as shown in Figure 4.6. Marker position measurements were filtered to 250 Hz using a low-pass smoother.

To compute the state vector, rigid bodies were fit to the marker data at each data frame. A simplex search algorithm was used find the center of mass position and orientation that minimized the cost function

$$J(\phi) = \frac{1}{2}\mathbf{e}^T\mathbf{W}\mathbf{e} \quad (4.6)$$

Figure 4.5 **Flight testing experimental setup for wing aerodynamics modeling**

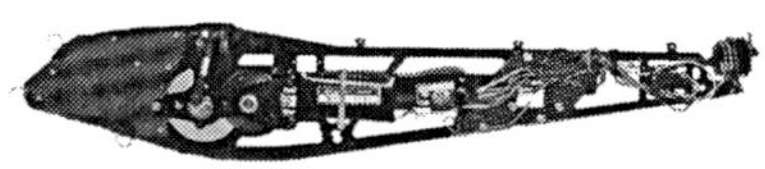

(a) fuselage

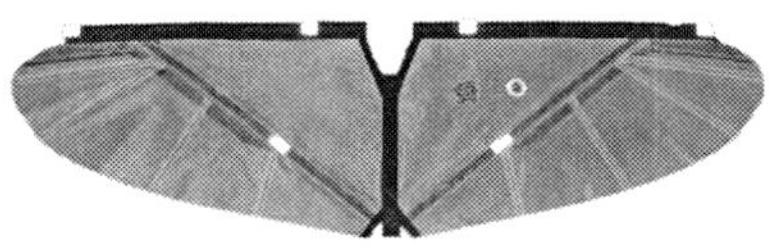

(b) wings

(c) tail

Figure 4.6 **Retro-reflective marker locations on the ornithopter**

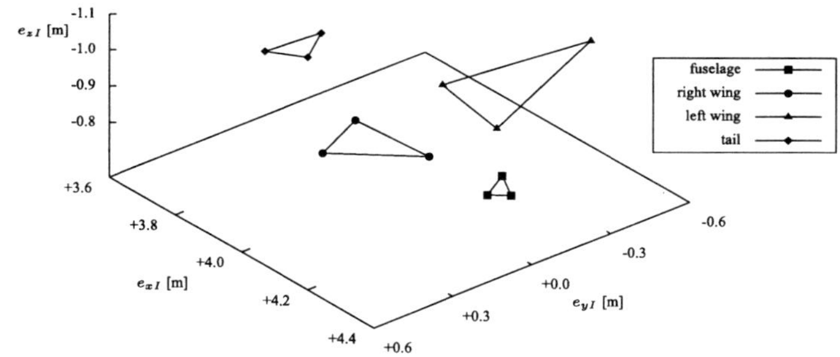

Figure 4.7 **Rigid body ornithopter geometry fit to retro-reflective marker data**

where ϕ is a vector containing the rigid body position and orientation quaternion, $\mathbf{W}$ is a unity weighting matrix, and the error

$$\mathbf{e} = \mathbf{r}_{k,I}^{I} - (\mathbf{r}_{ij,I}^{I} + \mathbf{R}^{I,ij}\mathbf{r}_{k,ij}^{ij}) \tag{4.7}$$

is the difference between the measured marker locations and those computed from the rigid body center of mass position, a rotation matrix derived from its orientation, and known positions of the markers relative to the center of mass. An example of the fit is shown in Figure 4.7. Errors for the marker measurements, as well as the rigid body positions and orientations, are given in Table 4.2. From these estimates, the position states are assembled and subsequently smoothly differentiated [37] to yield the velocity states. Previous studies have shown these measurements contain lower noise levels than measurements obtained using MEMS gyroscopes and accelerometers [72].

4.3.2 Measurements

Longitudinal state variables for a representative segment of flight data, where the ornithopter was flapping at 5.91 Hz, is shown in Figure 4.8. The periodic wing forcing induces a heave velocity between −1.1 m/s and +0.90 m/s, which results in a 0.08 m oscillation about the nominal altitude. Smoothly differentiating the heave velocity shows that at the

Table 4.2 **Marker position, rigid body position, and rigid body orientation standard errors**

Rigid Body	Marker	Position	Orientation
C_{ij}	$\sigma(\mathbf{r}^I_{k,I})$ [m]	$\sigma(\hat{\mathbf{r}}^I_{ij,I})$ [m]	$\sigma(\hat{\eta}^I_{ij,I})$ [rad]
fuselage	0.0009	0.0009	0.0056
right wing	0.0017	0.0011	0.0041
left wing	0.0018	0.0013	0.0037
tail	0.0008	0.0006	0.0036

fuselage center of mass, the ornithopter experiences translational accelerations between −4.0 and +4.9 times the that of gravity. The wing beats also induce a pitch rate between −1.4 and +1.6 rad/s, resulting in a 0.02 rad oscillation about the mean pitch angle. These observations are consistent with the inertial measurements presented in Chapter 2. The longitudinal tail deflection, although commanded to a trim condition, oscillates due to the vehicle and actuator dynamics. The down stroke phase of the flapping cycle has a range of −0.67 to +0.39 rad, which occurs between $0.25T_f$ and $0.78T_f$, yielding a 1.18 down stroke to up stroke ratio.

Using the identified model structure (3.18), the flight data can be partitioned into contributions arising from inertial, dynamic coupling, gravitational, and aerodynamic sources. Magnitudes of force and moment contributions are shown in this manner in Figure 4.9 as polar plots, where the angle represents the location within the flapping cycle and the radius denotes the magnitude. Figure 4.9(a) shows that forces arising from inertial and wing aerodynamic contributions dominate just before and after transitions between up and down stroke. Figure 4.9(b) shows that the nonlinear coupling and wing aerodynamic terms contribute most heavily to the moments throughout the wing stroke, but diminish at the transitions between up and down stroke, which is when the wing changes direction.

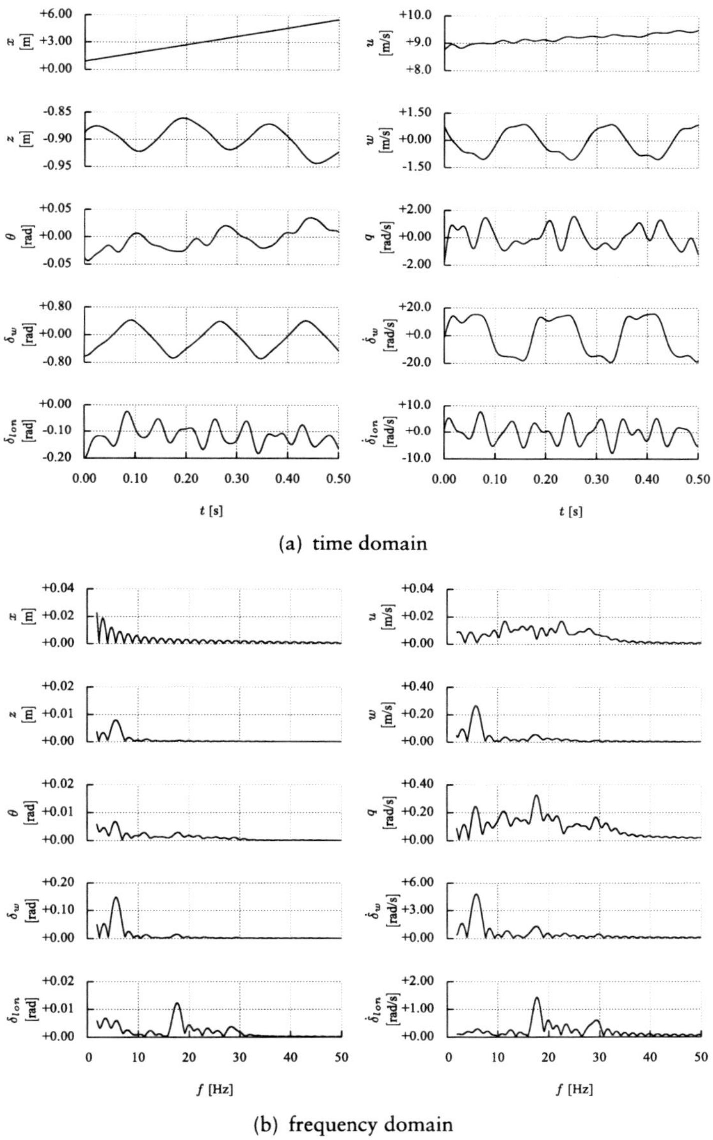

Figure 4.8 **Longitudinal state variable measurements from flight data**

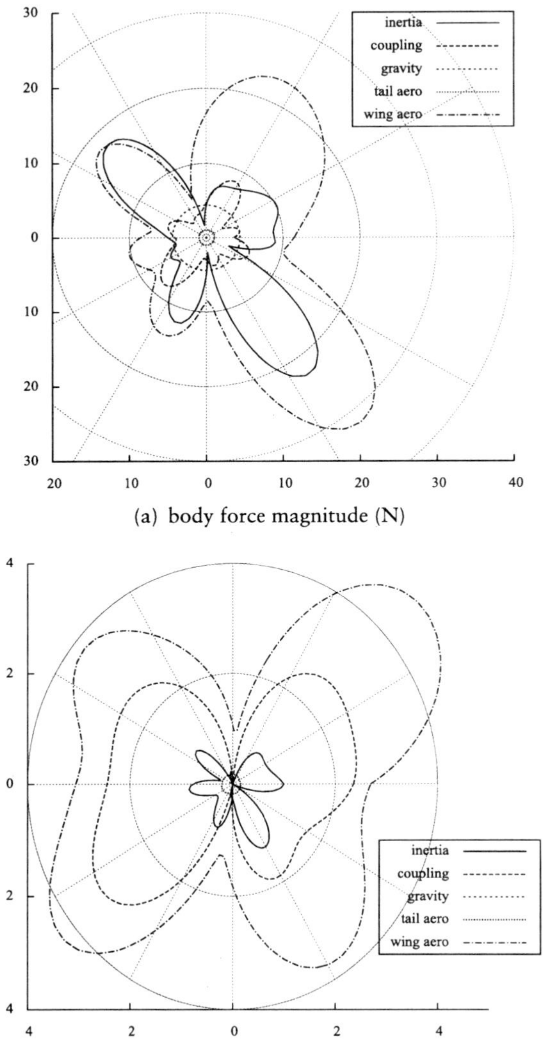

(a) body force magnitude (N)

(b) body moment magnitude (Nm)

Figure 4.9 **Force and moment magnitudes over a wing stroke**

In a similar fashion, kinematic quantities can be displayed through the wing stroke, as in Figure 4.10, where the angular coordinate denotes the position in the flapping cycle, and where the radial component denotes the position along the wing span. Figure 4.10(a) shows estimates of the main wing spar bending deformation, found by interpolating three marker locations along the wing span. Maximum tip deflections of ± 0.02 m, which are relatively small, are found $0.13\mathcal{T}_f$ after the transitions between up and down strokes. Figure 4.10(b) shows the velocity magnitude, which has a minimum value of 8.9 m/s due to forward velocity, and a maximum value at the wing tips of 11.0 m/s due to the flapping motion. The angle of attack distribution is shown in Figure 4.10(c), which varies between -0.94 rad and $+0.86$ rad. The Reynolds number is shown in Figure 4.10(d) and varies between 19,000 and 232,000 for this flight, which encompasses the transition region where steady lift to drag ratios decrease [4]. Reduced frequency measurements in Figure 4.10(e) show that this flow regime may be classified as unsteady [18].

4.3.3 Results

System identification was performed to determine a model for the longitudinal force, heave force, and pitching moment generated by the wings. Rearranging (3.18), the generalized aerodynamic force due to each wing is

$$\mathbf{a}_{wing}(\mathbf{p},\mathbf{v}) = \tau - \left[\boldsymbol{M(p)}\dot{\boldsymbol{v}} + \mathbf{C}(\mathbf{p},\mathbf{v})\mathbf{v} + \mathbf{g}(\mathbf{p}) + \mathbf{a}_{tail}(\mathbf{p},\mathbf{v})\right] \tag{4.8}$$

where $\mathbf{a}_{tail}(\mathbf{p},\mathbf{v})$ is the aerodynamic contributions of the tail identified in the previous section. The first, third, and fifth elements of this vector represent the longitudinal force, heave force, and pitching moment, respectively, and are normalized in a similar fashion as (4.3). Other forces and moments were not identified because the excitation during flight testing was only in the longitudinal axis.

Model structure determination was performed in the time domain with equation-error and step-wise regression. The

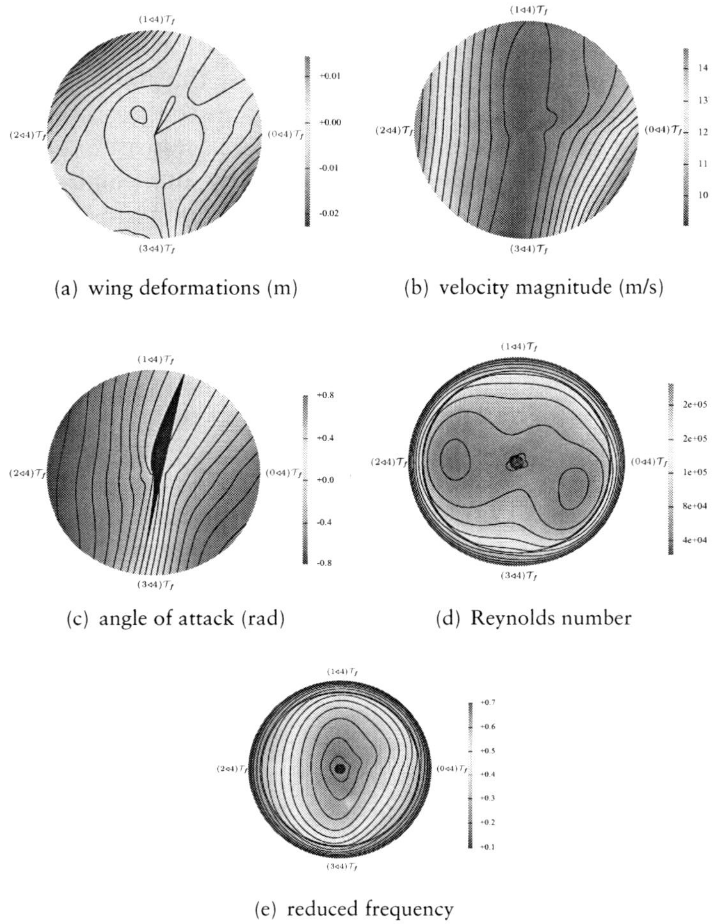

(a) wing deformations (m) (b) velocity magnitude (m/s)

(c) angle of attack (rad) (d) Reynolds number

(e) reduced frequency

Figure 4.10 **Scalar measurement distributions over a wing stroke**

resulting model structure was

$$
\begin{aligned}
C_x &= C_{x_0} + C_{x_{\delta_w}}\delta_w + C_{x_q}q + C_{x_{\dot{\delta}_w}}\dot{\delta}_w + C_{x_{\delta_w^2}}\delta_w^2 \\
C_z &= C_{z_0} + C_{z_q}q + C_{z_{\dot{\delta}_w}}\dot{\delta}_w \\
C_m &= C_{m_0} + C_{m_{\dot{\delta}_w}}\dot{\delta}_w
\end{aligned} \qquad (4.9)
$$

which is a nonlinear expansion in terms of the state variables. Model structures were also found using orthogonal regressors [73] and a traditional quasi-steady model [36], but yielded less accurate results. The thrust force includes a polar to capture the effects of drag, and also depends on the pitch rate and wing motion. The lift is a function of the pitch rate and wing motion, similar to a quasi-steady model. The pitching moment variation is due only to the wing motion.

As many state variables are periodic with harmonics of the flapping frequency, a collinearity analysis was performed on the model regressors. One method examined the correlation matrix between the regressors, which had maximum value of 0.57; because this value is less than 0.9, guidelines indicate that collinearity is not an issue [37, 74]. Additionally, an Eigensystem analysis was performed, which resulted in condition numbers under 3.70; because these numbers are much less than 100, guidelines also indicate here that collinearity is not an issue [37, 75].

Equation-error was used to find the parameters in (4.9) that fit the aerodynamic coefficients found using (4.8). Time domain and frequency domain fits are shown in Figure 4.11 and parameter estimates are provided in Table 4.3. The time domain results for the longitudinal force coefficient had larger error bounds and a lower fit value than the frequency domain results. While frequency domain modeling is often superior to time domain, this may also be due to low excitation, a short data record, or a large number of regressors. Heave force results were similar to the longitudinal force results. Moment coefficient results were very similar and accurate.

The identified models were used to predict the aerodynamic forces and moments in a second set of flight data. These results are shown in Figure 4.12. The heave force and the pitching moment were well predicted by both models. The longitudinal force model, however, appears to be over-parametrized; longer data records with more airspeed

Table 4.3 Wing aerodynamic parameters and standard errors

Parameter ϕ	Time Estimate $\hat{\phi} \pm \sigma(\hat{\phi})$	Frequency Estimate $\hat{\phi} \pm \sigma(\hat{\phi})$	Time Fit R^2	Frequency Fit R^2
C_{x_0}	$+0.2262 \pm 0.1068$	$+0.0634 \pm 0.0145$		
$C_{x_{\delta_w}}$	-0.1010 ± 0.0896	-0.0568 ± 0.0071		
C_{x_q}	-0.1913 ± 0.1168	-0.3647 ± 0.0068	0.66	0.80
$C_{x_{\dot{\delta}_w}}$	$+0.0127 \pm 0.0030$	$+0.0205 \pm 0.0002$		
$C_{x_{\delta_w^2}}$	-1.4455 ± 0.2236	-1.1843 ± 0.0165		
C_{z_0}	-0.9222 ± 0.2052	-0.9655 ± 0.0628		
C_{z_q}	-1.8822 ± 0.4566	-2.4701 ± 0.0594	0.73	0.86
$C_{z_{\dot{\delta}_w}}$	-0.0627 ± 0.0197	-0.0619 ± 0.0018		
C_{m_0}	$+0.0955 \pm 0.0172$	$+0.0942 \pm 0.0154$	0.96	0.96
$C_{m_{\dot{\delta}_w}}$	$+0.0862 \pm 0.0038$	$+0.0836 \pm 0.0005$		

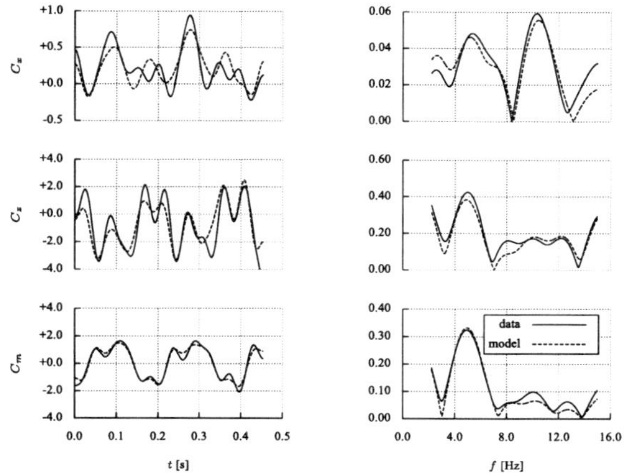

Figure 4.11 **Parameter estimation of wing aerodynamic models using equation-error in the time and frequency domains**

excitation, e.g. from flap variation or during push-over/pull-up maneuvers, are required to improve this model. The frequency domain model had higher fit values and was selected as the wing aerodynamic model.

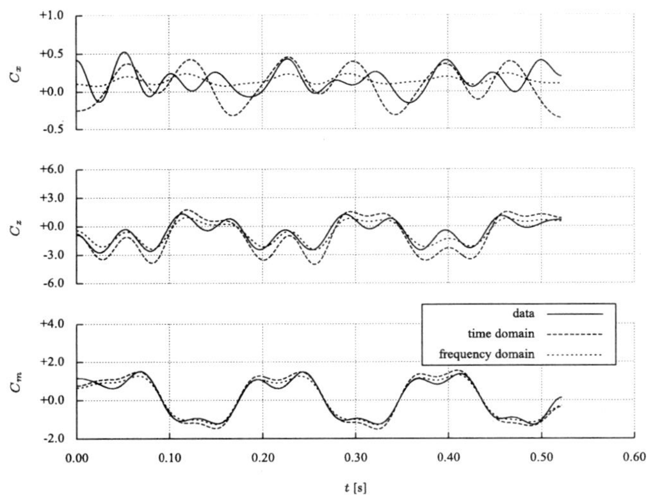

Figure 4.12 **Wing aerodynamic model prediction case**

4.4 Chapter summary

Previously in Chapter 3, the rigid body dynamics of the multibody ornithopter system were derived analytically. This chapter applied system identification techniques to extract models for the aerodynamic forces on the tail and wings from experimental data.

A wind tunnel test was conducted to determine the aerodynamic forces resulting from deflections of the tail flap. Simple models based on angle of attack and side slip angle were able to describe the aerodynamic forces generated.

A flight test was conducted to observe the ornithopter under free flight conditions. A visual tracking system was employed to measure the spatial position of retro-reflective markers placed on the ornithopter, from which the state vector was computed. Several kinematic variables were presented for the ornithopter over the course of a wing stroke from flight data, providing insight into the flight physics. System identification yielded a low-order model for the aerodynamic forces generated by the flapping wings.

This chapter and the previous chapter fully describe the nonlinear, multibody model of the ornithopter flight dynamics, determined using a combination of analytical and experimental methods.

5

Simulation results

DOI: 10.1533/9780857094674.71

Abstract: In the previous chapters, the equations of motion were determined for the ornithopter. In this chapter, those equations are programmed into a nonlinear flight simulator for the purposes of simulation and model simplification. An overview of the simulation environment is first presented. Afterwards, a method is discussed for trimming the model for straight and level mean flight. Numerical methods are then employed to numerically linearize the model about straight and level mean flight, resulting in both a conventional time-invariant model as well as a time-periodic model.

Key Words: numerical linearization, linear time-invariant, linear time-periodic

5.1 Software simulation architecture

The ornithopter flight dynamics (3.7) and (3.18) were written as the first-order state space model

$$\dot{z} = \mathbf{f}(\mathbf{z}, \mu) \qquad (5.1)$$

and were coded in both FORTRAN 95 and MATLAB, according to the block diagram shown in Figure 5.1. The FORTRAN code ran very quickly; however, at the expense of computation time, the MATLAB environment provided a wealth of internal functions and tools such as plotting and optimization routines. Simulation of a single wing flap using MATLAB used approximately 0.94 seconds of computation time on a dual-core laptop, which was sufficient for this work.

A main script is first run, which initializes an ornithopter configuration file, simulation parameters, control laws, and an initial state vector. This information is given to an ordinary differential equation (ODE) solver, and returns the state vector time history. Data is then plotted, animated, analyzed, and saved. The ODE solver numerically integrates the equations of motion from an initial state and returns the state vector time history. In the FORTRAN implementation of the code, an explicit, fourth-order, Runge-Kutta method was employed, where inversion of the mass matrix was accomplished using a Cholesky decomposition and back-substitution. Solution time histories converged using a 10^{-4} s fixed time step interval. The MATLAB implementation of the code utilized the internal solver `ode15s`, which was chosen because it is an adaptive time step method for accurate solutions, implicit solver for saving computation time, and valid for stiff ODE problems.

The ODE solver calls the equations of motion file and computes the current state derivative. Within the routine, several subroutines are called. The first is a configuration file, specific to each ornithopter, that loads mass and inertia properties, linkage lengths, and actuator constants into the workspace. Next is a kinematics subroutine, which computes the necessary rotation matrices and kinematic variables. Lastly, subroutines are called that assemble the mass matrix, coupling matrix, gravitational and aerodynamic forces, and control inputs.

Several checks were performed to validate the flight simulation. First, the hand-derived equations of motion were checked against output of the software algebra package Mathematica. Second, simulations from random initial conditions without external forces resulted in state trajectories having constant kinetic energy. Third, simulations from random initial conditions with only conservative forces resulted

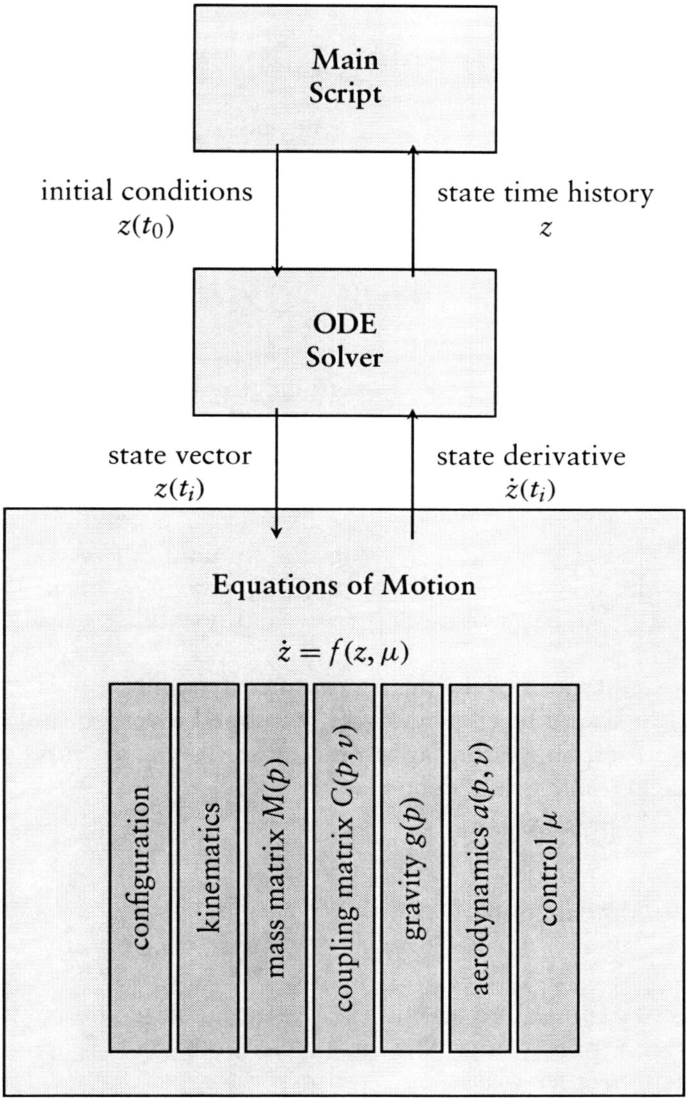

Figure 5.1 **Software simulation block diagram**

in state trajectories having constant total energy, again within numerical accuracy.

5.2 Determining trim solutions

Conventional aircraft have trim solutions characterized by fixed points in the state space that correspond to a constant state $\mathbf{z}^*$ and control μ^*. Nonlinear, iterative solvers are traditionally used to numerically solve

$$\dot{\boldsymbol{z}} = \mathrm{f}(\mathrm{z}^*, \boldsymbol{\mu}^*) = 0 \tag{5.2}$$

for the trim states and controls. Flapping-wing vehicles, on the other hand, have $\mathcal{T}_f$-periodic forcings which manifest as limit cycles, or trim conditions that are closed trajectories in the state space. Stable limit cycles are relatively easy to find because state trajectories beginning within the basin of attraction converge on the limit cycle as time progresses. While Lee et al. [26] developed a simulation that exhibits a stable limit cycle behavior, the majority of work in the literature consists of models that yield instabilities in forward flight [25, 50, 43, 76, 23].

Simulations of the the ornithopter dynamics model, during straight and level mean flight, exhibited divergent flight trajectories, suggesting an unstable limit cycle. As limit cycles are periodic orbits defined as

$$\boldsymbol{z}(t) = \boldsymbol{z}(t + \mathcal{T}_f) \tag{5.3}$$

the cost function

$$J(\phi) = \frac{1}{2}\mathbf{e}^T\mathbf{W}\mathbf{e} \tag{5.4}$$

can be minimized to find the the initial state and control to start a trajectory on this limit cycle. In this cost function, the error vector

$$\boldsymbol{e} = \begin{bmatrix} z(\mathcal{T}_f) - z(0) \\ \theta(\mathcal{T}_f) - \theta(0) \\ u(\mathcal{T}_f) - u(0) \\ w(\mathcal{T}_f) - w(0) \\ q(\mathcal{T}_f) - q(0) \end{bmatrix} \tag{5.5}$$

is the difference between the longitudinal states after one flap cycle, and the unknown parameter vector

$$\phi = \begin{bmatrix} \theta(0) \\ u(0) \\ w(0) \\ q(0) \\ \delta^*_{lon} \end{bmatrix} \quad (5.6)$$

consists of the initial longitudinal states and longitudinal trim control setting. The simplex search optimization method was employed to minimize the cost function (5.4), where the weighting matrix

$$\mathbf{W} = diag\{1, 10000, 10, 100, 100\} \quad (5.7)$$

accounted for the relative magnitudes in the error vector and produced adequate results. Figure 5.2 shows an example of the errors and the total cost for a solution, which required approximately 200 iterations before convergence, although solutions were close after about 50 iterations. This is a simple method to trim to the ornithopter; other more complicated methods in the literature include multiple shooting algorithms [25] and monitoring mean values of flight variables after an initial launching force vector is supplied [28].

The ornithopter was trimmed for a range of flapping frequencies between 4 Hz and 10 Hz. Below this range the ornithopter could not be trimmed, and above this range the dynamics model loses validity. Flight variables are shown as a function of flapping frequency in Figure 5.3. The longitudinal tail deflection becomes more negative as the wings flap faster in order to counter an increasing pitching moment. Although this deflection adds more drag, even more thrust is produced as the flapping rate increases, which in turn increases the mean forward speed. The faster wing movements also generate a lower mean pitch angle and oscillation, as the inertial effects cause a more rapid pitching motion that has less time to travel. Interestingly, the pitch rate amplitude remains approximately constant with flapping frequency because the tail is adjusted to compensate for the mean change in the pitching moment, which has also been reported by Lee et al. [28]. In addition, the altitude oscillation decreases and the heave velocity oscillation increases with increasing flapping frequency. Similar to the

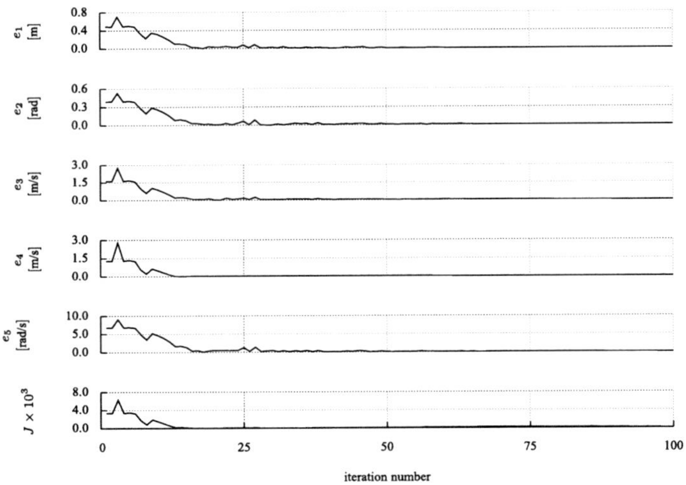

Figure 5.2 **Optimization results for finding limit cycle trim trajectories**

pitching dynamics, this is due to a faster excitation, behind which the response lags.

The main features of the trim condition consist of the heave dynamics, involving the altitude and heave velocity, and the pitch dynamics, involving the pitch angle and pitch rate, which are shown in Figure 5.4. The characteristics shown in Figure 5.3 are evident in these plots, as well as a change in shape, or bifurcation, of the limit cycles around 7 Hz. A similar phenomenon has been reported by Lee et al. [28], who claimed that simplified models may be possible for these higher flapping frequencies.

The remainder of this chapter will deal with a single flapping frequency of 5.91 Hz, to correspond with the flight test used for system identification of the aerodynamics model. The trimmed flight trajectory for this flapping frequency is shown in Figure 5.5, which has similar trajectories as those reported in the literature [51, 28]. Due to the instability of the trim solution, integration over more than approximately five wing strokes will tend to diverge from the limit cycle due to the accumulation of numerical errors. The trimmed

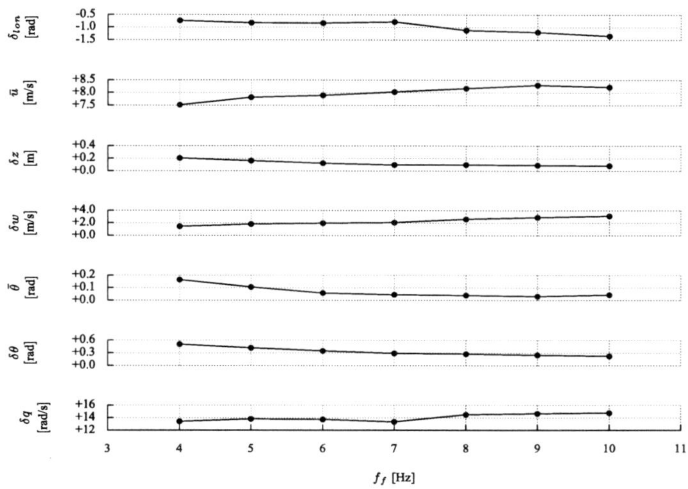

Figure 5.3 **Variations in trim characteristics with flapping frequency**

model resulted in lower longitudinal velocity mean and oscillations than measured in the visual tracking flight test, which is due to the difficulty in modeling the thrust and drag of the wings. The altitude and heave oscillations, as well as the pitch oscillations and rates were consistent with those measured in flight tests. The simulation data also has much lower frequency content than the flight data, due to the unmodeled fluid/structure interactions, actuator dynamics, and simplified aerodynamics.

5.3 Numerical linearization about straight and level mean flight

Within the range of its validity, a linearized model of the flight dynamics facilitates the use of modern linear analysis tools, which can provide physical insight into the system and be used to design control laws to guarantee performance metrics. Additionally, linear models can be incorporated with

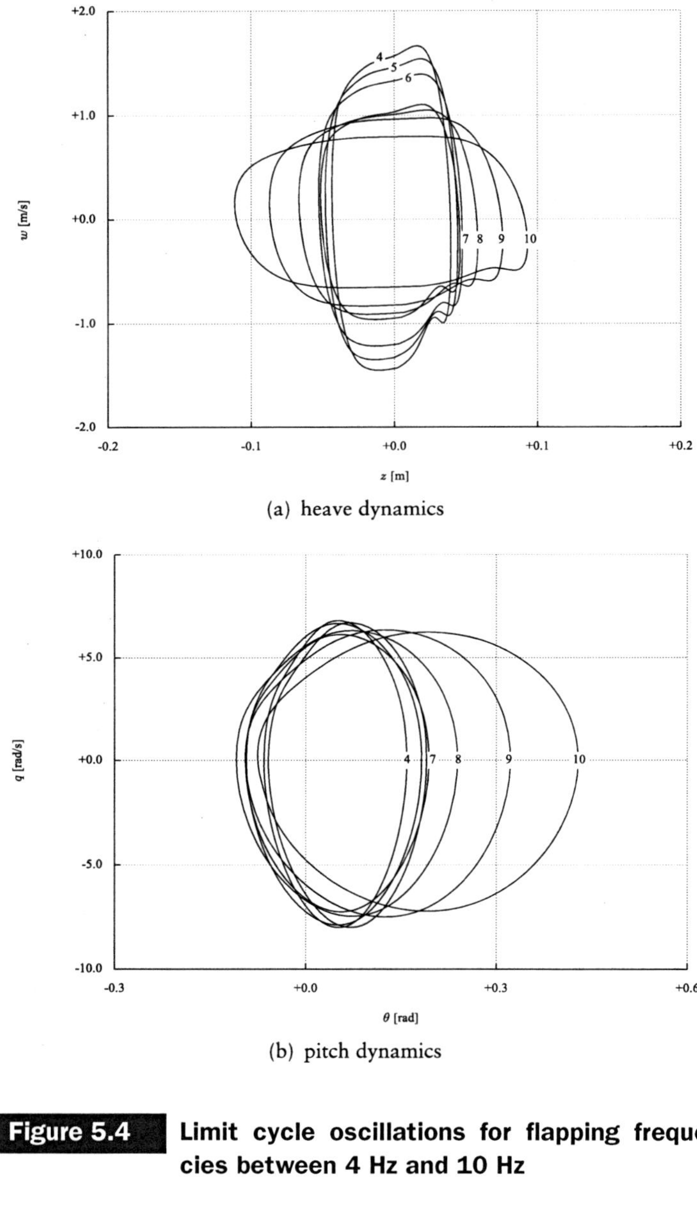

(a) heave dynamics

(b) pitch dynamics

Figure 5.4 **Limit cycle oscillations for flapping frequencies between 4 Hz and 10 Hz**

standard commercial autopilots. Towards these goals, a linear model of the ornithopter flight dynamics

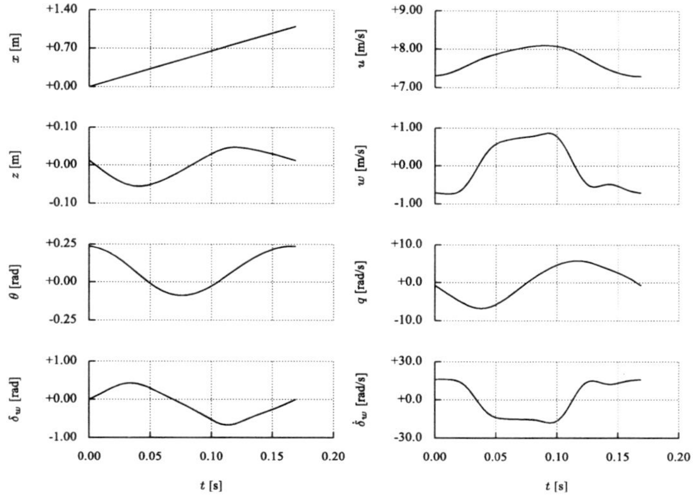

Figure 5.5 **Trimmed flight trajectory solution**

$$\dot{\boldsymbol{x}} = \mathbf{A}\mathbf{x} + \mathbf{B}\mathbf{u} \tag{5.8}$$

is desired, where

$$\begin{aligned} \mathbf{x} &= \mathbf{z} - \mathbf{z}^* \\ \mathbf{u} &= \boldsymbol{\mu} - \boldsymbol{\mu}^* \end{aligned} \tag{5.9}$$

are state and control perturbations away from the trim values. To parallel the models currently employed in autopilots, the reduced state and control vectors are

$$\mathbf{z} = \begin{bmatrix} \theta \\ u \\ w \\ q \\ \phi \\ v \\ p \\ r \end{bmatrix} \qquad \boldsymbol{\mu} = \begin{bmatrix} \delta_{lon} \\ \delta_{lat} \end{bmatrix}. \tag{5.10}$$

Linear models may be determined using a variety of methods. The nonlinear model may be linearized analytically using Jacobian matrices, but this method is prohibitive for complex and large-order systems. Alternatively, system identification experiments can be performed on the nonlinear simulation, but this method may require several iterations or long records of data to obtain adequate information content. In this work, the nonlinear model is linearized numerically, using finite differences, which requires little analysis and can be automated. A nonlinear system (5.1) can be numerically linearized about some point or trajectory in the state space $\mathbf{z}^*$ and some control μ^*. Let

$$\begin{aligned} \xi_1 &= \mathbf{f}(\mathbf{z}^* + \epsilon_i \mathbf{e}_i, \boldsymbol{\mu}^*) \\ \xi_2 &= \mathbf{f}(\mathbf{z}^* - \epsilon_i \mathbf{e}_i, \boldsymbol{\mu}^*) \\ \xi_3 &= \mathbf{f}(\mathbf{z}^* + 2\epsilon_i \mathbf{e}_i, \boldsymbol{\mu}^*) \\ \xi_4 &= \mathbf{f}(\mathbf{z}^* - 2\epsilon_i \mathbf{e}_i, \boldsymbol{\mu}^*) \end{aligned} \tag{5.11}$$

where ϵ_i is a small perturbation and $\mathbf{e}_i$ is an elementary vector. Then the ith column of the system state matrix $\mathbf{A}$ is computed

$$\frac{8(\xi_1 - \xi_2) - (\xi_3 - \xi_4)}{12\epsilon_i} \tag{5.12}$$

and is accurate to $O(\epsilon_i^3)$. A similar process, where instead each control variable is perturbed, is performed to find columns of the system control matrix $\mathbf{B}$. The perturbations ϵ_i are typically decreased iteratively until the system matrices converge [2].

5.3.1 Linear time-invariant model

Conventional aircraft flight dynamics models used for control design are typically linear time-invariant (LTI) models, where the system matrices are constant in time. The ornithopter flight dynamics can be cast into an LTI model by using averaging theory [77, 40, 38, 23]. The cycle-averaged state derivatives (5.11) are then

$$\xi_1 = (1/T_f) \int_0^{T_f} \mathbf{f}(\mathbf{z}^* + \epsilon_i \mathbf{e}_i, \boldsymbol{\mu}^*) dt$$

$$\begin{aligned}
\xi_2 &= (1/T_f)\int_0^{T_f} \mathrm{f}(\mathbf{z}^* - \epsilon_i \mathbf{e}_i, \boldsymbol{\mu}^*)dt \\
\xi_3 &= (1/T_f)\int_0^{T_f} \mathrm{f}(\mathbf{z}^* + 2\epsilon_i \mathbf{e}_i, \boldsymbol{\mu}^*)dt \\
\xi_4 &= (1/T_f)\int_0^{T_f} \mathrm{f}(\mathbf{z}^* - 2\epsilon_i \mathbf{e}_i, \boldsymbol{\mu}^*)dt \qquad (5.13)
\end{aligned}$$

and the remainder of the process is the same. Typically when this technique is applied, a time scale separation exists in the dynamics. Since the flapping frequency is close to the expected rigid body modes of the ornithopter, this method is questionable, as is the practice of using commercial autopilots, based on LTI models, with an ornithopter. The LTI model resulting from numerical linearization with averaging is

$$\begin{pmatrix} \Delta\dot{\theta} \\ \Delta\dot{u} \\ \Delta\dot{w} \\ \Delta\dot{q} \\ \Delta\dot{\phi} \\ \Delta\dot{v} \\ \Delta\dot{p} \\ \Delta\dot{r} \end{pmatrix} =$$

$$\begin{pmatrix}
-0.69 & +0.01 & -0.83 & +0.84 & -0.04 & -0.01 & +0.06 & -0.03 \\
-7.38 & -0.51 & +2.42 & -2.59 & +0.06 & -0.01 & -0.21 & +0.09 \\
+0.33 & -0.72 & -0.59 & +0.41 & +0.04 & +0.03 & -0.01 & -0.03 \\
-2.64 & -2.03 & -12.4 & +0.74 & +0.26 & +0.12 & +0.66 & -0.62 \\
-0.00 & -0.00 & -0.00 & -0.00 & -0.50 & -0.12 & +0.87 & +0.42 \\
-0.00 & +0.00 & -0.00 & -0.01 & +8.65 & -0.60 & +1.65 & -6.27 \\
+0.00 & -0.01 & -0.08 & +0.00 & +4.79 & +1.87 & -0.14 & -3.27 \\
-0.01 & +0.01 & -0.00 & -0.00 & -0.13 & +0.06 & +1.67 & +0.49
\end{pmatrix}$$

$$\times \begin{pmatrix} \Delta\theta \\ \Delta u \\ \Delta w \\ \Delta q \\ \Delta\phi \\ \Delta v \\ \Delta p \\ \Delta r \end{pmatrix} + \begin{pmatrix} -3.10 & +0.27 \\ +10.0 & -0.82 \\ -2.09 & +0.32 \\ -44.5 & +3.37 \\ -0.02 & +0.05 \\ -0.03 & -3.20 \\ -0.30 & -2.91 \\ -0.02 & +4.84 \end{pmatrix} \begin{pmatrix} \Delta\delta_{lon} \\ \Delta\delta_{lat} \end{pmatrix}. \qquad (5.14)$$

where states represent perturbations from an averaged trim value. While linearized lateral dynamics are presented in this chapter for completeness, their validity is questionable, as lateral motions were not excited during the flight test used to model the wing aerodynamics in Chapter 4.

There are several points of interest in these matrices as compared to those of a fixed-wing aircraft, for example the F-16 model in Appendix B.3. This ordering of the state vector illustrates that the state matrix decouples in much the same manner as conventional aircraft in straight and level flight. The first and second 4×4 block-diagonal matrices represents the longitudinal and lateral dynamics, respectively. Here however, the pitch rate dynamics however still have significant coupling with the lateral dynamics, which is not common to fixed-wing aircraft and is due to the wings flapping. The block-diagonal matrices are also more heavily populated than models for fixed-wing aircraft: typically the equations for the pitch angle and bank angle derivatives involve only the pitch rate and roll rate; however, for the ornithopter, they involve many states. The equations for the forward and side velocity typically involve the acceleration of gravity; here the signs are correct but the value is slightly lower than expected.

The control matrix is also significantly populated and decouples in a similar manner as the state matrix. A positive longitudinal tail angle perturbation creates less drag and negative lift on the tail, resulting in a positive forward acceleration, a negative heave acceleration, and a negative pitching moment. A positive lateral tail angle perturbation creates a negative side force, a negative roll rate, and a positive yaw rate. The controllability matrix for this system has full rank and is fully controllable.

The modes of this system, described by Eigenvalues and Eigenvectors of the state matrix, are shown in Figures 5.6 and 5.7 and tabulated in Table 5.1. The longitudinal dynamics include two oscillatory modes, one unstable and one stable, occurring at approximately the same frequency. The unstable mode is under-damped, while the stable mode is highly damped. The Eigenvectors of these modes indicate that they both are short period type motions, involving a

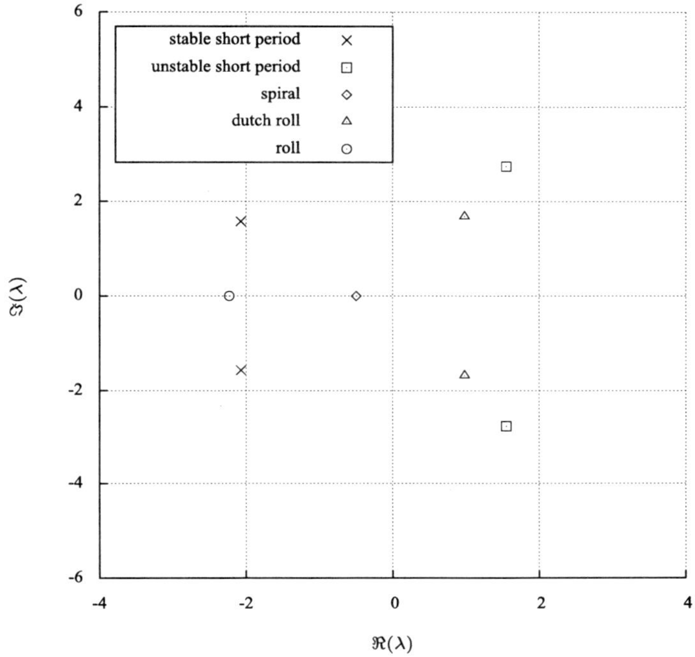

Figure 5.6 **Linear time-invariant model pole locations**

fast and strong interplay of the pitch rate and forward velocity. Unlike conventional aircraft, a phugoid mode does not appear. The classical spiral, dutch roll, and roll modes are present in the decoupled lateral dynamics. The spiral mode is first order and unstable, involving mostly the sidewards velocity. The dutch roll mode is unstable and is strongly dependent on the sideways velocity. The roll mode is a first order mode, strongly dependent on the roll rate.

5.3.2 Linear time-periodic model

There is not a time scale separation between the periodicity of the wings flapping and the rigid body modes found in the LTI model, which calls into question the validity of using LTI models. A linear time-periodic (LTP) model results from

Table 5.1 **Modal parameters of the decoupled linear time-invariant model**

Mode	Eigenvalue λ_i	Eigenvector $\mathbf{v}_i$	Damping Ratio ζ_i	Frequency [rad/s] ω_i
stable short period	$-2.07 \pm 1.57j$	$\begin{bmatrix} +0.17 \pm 0.19j \\ -0.56 \pm 0.32j \\ -0.11 \mp 0.00j \\ -0.72 \end{bmatrix}$	+0.80	2.60
unstable short period	$+1.55 \pm 2.75j$	$\begin{bmatrix} -0.14 \mp 0.06j \\ +0.74 \\ -0.18 \pm 0.13j \\ -0.36 \mp 0.50j \end{bmatrix}$	−0.49	3.16
spiral	−0.50	$\begin{bmatrix} +0.58 \\ -0.18 \\ -0.36 \\ +0.71 \end{bmatrix}$	+1.00	0.50
dutch roll	$+0.98 \pm 1.68j$	$\begin{bmatrix} +0.07 \mp 0.25j \\ +0.49 \pm 0.23j \\ +0.58 \\ +0.18 \mp 0.53j \end{bmatrix}$	−0.51	1.94
roll	−2.23	$\begin{bmatrix} +0.26 \\ +0.65 \\ -0.61 \\ +0.37 \end{bmatrix}$	+1.00	2.23

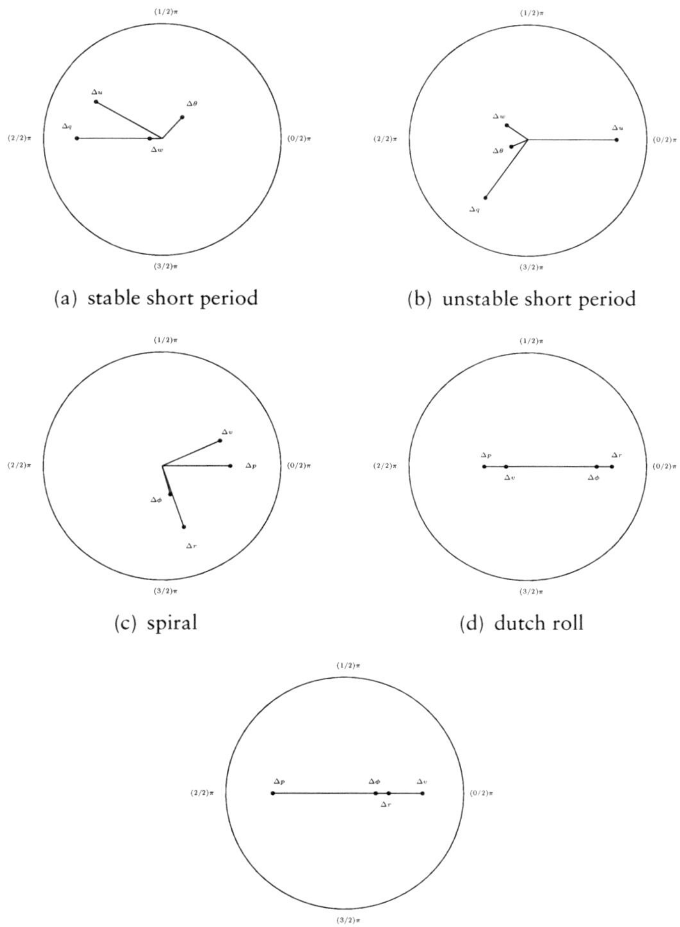

(a) stable short period

(b) unstable short period

(c) spiral

(d) dutch roll

(e) roll

Figure 5.7 **Linear time-invariant model Eigenvector polar plots**

computing (5.11) at each time step over the wing stroke, so that the system matrices vary with time. Elements of the system matrices are shown for the longitudinal and lateral decoupled systems in Figures 5.3.2 and 5.3.2, respectively.

From observations of the frequency content, a functional representation of the matrix elements was obtained using the model

$$y = \phi_0 + \phi_1 \sin(2\pi f_f t) + \phi_2 \sin(4\pi f_f t) + \phi_3 \cos(2\pi f_f t) \quad (5.15)$$

and applying an equation-error analysis, where $\{\phi\}_i$ are unknown, constant coefficients. The functions consists of a bias and harmonic functions that are T_f-periodic and capture the major features of the responses. Simulation data and curve fits are shown in Figure 5.8, whereas parameter estimates, standard error bounds, and coefficients of determination for each matrix element are provided in Table 5.2. Generally the function (5.15) fits all the elements well with high coefficients of determination and relatively low standard errors. Whereas the curve fits for the lateral dynamics retain most of the frequency content, some elements in the longitudinal model, for instance a_{41} and a_{43}, contain frequency content up to 10 times the flapping frequency and is not captured by the curve fit. This higher frequency content in the longitudinal dynamics is due to the longitudinal flapping of the wings.

For a linear time varying system with $\mathbf{x} \in \Re^{n\times 1}$, controllability at time t requires that the matrix

$$\mathcal{C}(t) = \left[\begin{array}{cccc} \mathbf{M}_0(t) & \mathbf{M}_1(t) & \ldots & \mathbf{M}_{n-1}(t) \end{array} \right] \quad (5.16)$$

have rank n, where in general

$$\mathbf{M}_{k+1}(t) = -\mathbf{A}(t)\mathbf{M}_k(t) + \dot{\mathbf{M}}_k(t) \quad (5.17)$$

$$\mathbf{M}_0(t) = \mathbf{B}(t) \quad (5.18)$$

for $k \in 0, n-2$. Employing (5.15) and the identified parameters, computation of the controllability matrix for a wing stroke shows that both the longitudinal and lateral decoupled dynamics are fully controllable for all time.

Given the functional representation of the LTP system (5.15), the state transition matrix can be computed by assembling the responses to initial conditions $\mathbf{x}_i(0) = \mathbf{e}_i$ as a fundamental matrix

Table 5.2 **Estimated parameters and standard errors for the decoupled linear time-periodic model**

element	$\hat{\phi}_0 \pm s(\hat{\phi}_0)$	$\hat{\phi}_1 + \pm s(\hat{\phi}_1)$	$\hat{\phi}_2 \pm s(\hat{\phi}_2)$	$\hat{\phi}_3 \pm s(\hat{\phi}_3)$	R^2
a_{11}	-0.44 ± 0.05	$+0.19 \pm 0.07$	$+0.46 \pm 0.07$	$+0.31 \pm 0.07$	0.28
a_{21}	-9.03 ± 0.10	-2.31 ± 0.15	-0.52 ± 0.15	-0.25 ± 0.14	0.62
a_{31}	-0.18 ± 0.30	-0.70 ± 0.43	-1.23 ± 0.43	-0.67 ± 0.43	0.07
a_{41}	-18.7 ± 1.71	$+4.23 \pm 2.43$	$+12.8 \pm 2.43$	$+14.0 \pm 2.42$	0.28
a_{12}	$+0.13 \pm 0.02$	-0.62 ± 0.03	$+0.05 \pm 0.03$	-0.44 ± 0.02	0.85
a_{22}	-1.24 ± 0.07	$+4.91 \pm 0.10$	-0.83 ± 0.10	$+1.75 \pm 0.10$	0.95
a_{32}	-0.39 ± 0.24	$+7.63 \pm 0.34$	-2.23 ± 0.34	$+0.86 \pm 0.34$	0.77
a_{42}	$+1.89 \pm 1.62$	$+14.3 \pm 2.29$	-17.1 ± 2.29	-27.5 ± 2.28	0.59
a_{13}	$+0.21 \pm 0.02$	-0.40 ± 0.03	$+0.34 \pm 0.03$	-0.07 ± 0.03	0.66
a_{23}	-1.53 ± 0.09	$+8.02 \pm 0.12$	-1.83 ± 0.12	$+3.03 \pm 0.12$	0.97
a_{33}	-0.12 ± 0.27	$+3.92 \pm 0.38$	-1.83 ± 0.38	$+0.03 \pm 0.38$	0.43
a_{43}	-11.3 ± 1.30	$+6.82 \pm 1.85$	-8.52 ± 1.85	-14.0 ± 1.84	0.36
a_{14}	$+0.79 \pm 0.00$	$+0.19 \pm 0.01$	-0.04 ± 0.01	$+0.12 \pm 0.01$	0.87
a_{24}	-1.14 ± 0.02	-0.60 ± 0.03	$+0.04 \pm 0.03$	-0.81 ± 0.03	0.87
a_{34}	$+0.35 \pm 0.02$	-0.90 ± 0.03	$+0.07 \pm 0.03$	$+0.51 \pm 0.03$	0.90
a_{44}	-5.50 ± 0.30	-0.21 ± 0.43	-2.10 ± 0.43	$+7.76 \pm 0.42$	0.68
b_1	$+0.55 \pm 0.02$	-0.90 ± 0.03	$+0.19 \pm 0.03$	-0.17 ± 0.03	0.88
b_2	-5.55 ± 0.09	$+3.71 \pm 0.13$	$+0.11 \pm 0.13$	$+3.69 \pm 0.12$	0.91
b_3	$+0.78 \pm 0.03$	$+0.77 \pm 0.04$	$+0.06 \pm 0.04$	-2.05 ± 0.04	0.94
b_4	-15.0 ± 0.86	-12.6 ± 1.22	$+15.2 \pm 1.22$	-36.5 ± 1.21	0.88
a_{11}	-0.90 ± 0.01	$+1.33 \pm 0.01$	-0.55 ± 0.01	$+0.66 \pm 0.01$	0.99
a_{21}	$+8.21 \pm 0.07$	$+1.99 \pm 0.09$	$+0.35 \pm 0.09$	$+0.17 \pm 0.09$	0.74
a_{31}	-28.0 ± 0.66	-0.61 ± 0.94	$+1.13 \pm 0.94$	$+33.5 \pm 0.93$	0.89
a_{41}	-1.43 ± 0.03	$+1.83 \pm 0.05$	-0.78 ± 0.05	$+1.45 \pm 0.05$	0.94
a_{12}	-0.22 ± 0.01	$+1.23 \pm 0.01$	-0.07 ± 0.01	$+0.49 \pm 0.01$	0.98
a_{22}	-0.76 ± 0.06	$+2.06 \pm 0.09$	$+0.49 \pm 0.09$	$+0.07 \pm 0.09$	0.78
a_{32}	$+4.41 \pm 0.71$	-19.6 ± 1.00	$+8.24 \pm 1.00$	$+46.6 \pm 1.00$	0.94
a_{42}	$+0.04 \pm 0.07$	$+1.30 \pm 0.10$	$+0.20 \pm 0.10$	$+0.96 \pm 0.10$	0.61
a_{13}	$+0.89 \pm 0.00$	$+0.27 \pm 0.00$	-0.06 ± 0.00	$+0.09 \pm 0.00$	1.00
a_{23}	$+1.24 \pm 0.02$	$+0.20 \pm 0.02$	$+0.15 \pm 0.02$	$+0.00 \pm 0.02$	0.45
a_{33}	-5.94 ± 0.12	$+1.00 \pm 0.18$	$+0.82 \pm 0.18$	$+8.62 \pm 0.18$	0.94
a_{43}	$+1.47 \pm 0.04$	-3.38 ± 0.06	$+0.59 \pm 0.06$	-0.61 ± 0.06	0.96
a_{14}	$+0.67 \pm 0.01$	-0.61 ± 0.01	$+0.06 \pm 0.01$	-0.36 ± 0.01	0.97
a_{24}	-5.99 ± 0.04	-1.16 ± 0.05	-0.58 ± 0.05	$+0.36 \pm 0.05$	0.80
a_{34}	$+20.9 \pm 0.43$	$+4.60 \pm 0.61$	$+0.02 \pm 0.61$	-22.6 ± 0.61	0.90
a_{44}	$+1.79 \pm 0.26$	-0.76 ± 0.36	$+4.68 \pm 0.36$	-3.30 ± 0.36	0.60
b_1	$+0.25 \pm 0.00$	-0.20 ± 0.01	$+0.01 \pm 0.01$	-0.12 ± 0.01	0.91
b_2	-0.24 ± 0.01	-0.26 ± 0.02	-0.26 ± 0.02	$+0.28 \pm 0.02$	0.85
b_3	$+9.27 \pm 0.13$	-0.36 ± 0.19	$+0.73 \pm 0.19$	-8.58 ± 0.19	0.93
b_4	$+6.64 \pm 0.12$	$+0.17 \pm 0.17$	$+2.46 \pm 0.17$	-1.95 ± 0.16	0.69

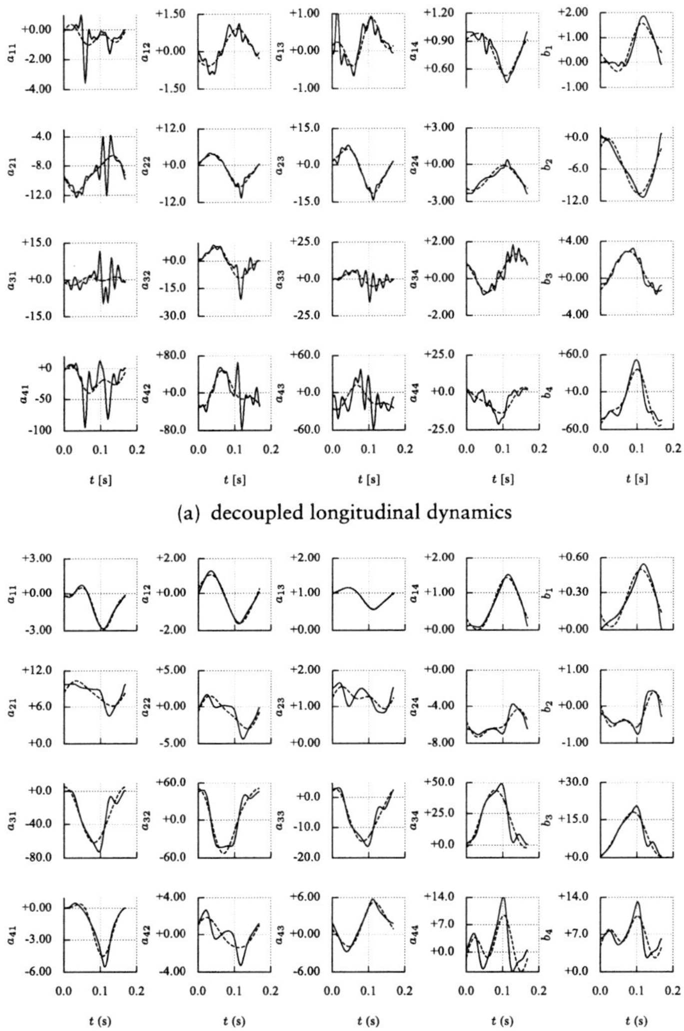

(a) decoupled longitudinal dynamics

(b) decoupled lateral dynamics

Figure 5.8 **Numerically linearized (solid) and curve fitted (dashed) decoupled system matrix element time histories over one wing stroke**

$$\mathbf{\Phi}(\mathcal{T}_f, 0) = \begin{bmatrix} \mathbf{x}_1(\mathcal{T}_f) & \mathbf{x}_2(\mathcal{T}_f) & \mathbf{x}_3(\mathcal{T}_f) & \mathbf{x}_4(\mathcal{T}_f) \end{bmatrix} \quad (5.19)$$

stemming from a Floquet decomposition [78]. The decoupled longitudinal and lateral systems result in the analogous LTI system matrices

$$\mathbf{A}_{lon} = \begin{bmatrix} +0.75 & +0.06 & -0.01 & +0.07 \\ -0.96 & +0.73 & -0.21 & -0.15 \\ +0.02 & +0.07 & +1.05 & +0.03 \\ -2.16 & +0.16 & -1.04 & +0.20 \end{bmatrix} \quad (5.20)$$

$$\mathbf{A}_{lat} = \begin{bmatrix} +0.53 & -0.06 & +0.05 & +0.40 \\ +1.05 & +1.00 & +0.21 & -0.84 \\ -1.95 & +0.69 & +0.25 & +1.48 \\ -1.20 & -0.38 & -0.10 & +2.08 \end{bmatrix} \quad (5.21)$$

from which the stability can be ascertained using a discrete Eigenvalue analysis. Poles are are tabulated in Table 5.3 and shown in Figure 5.9. The longitudinal dynamics exhibit a stable oscillatory mode, a stable subsidence mode, and an unstable divergence mode. The oscillatory mode looks most like a short period mode. The first order modes involve most of the state variables and in that regard are consistent with previous findings [43, 25, 51], although these modes are much slower. The lateral dynamics contain two subsidence modes and two divergence modes. One of the subsidence appears to be a roll mode, and one of the divergence appears as a spiral mode.

5.4 Modeling implications for control

In this chapter, three models have been presented: a nonlinear multibody model, a LTP model for straight and level mean flight, and a LTI model for straight and level mean flight.

The nonlinear model contains the full set of modeled dynamics, including the mass variation, inertial couplings, and nonlinear aerodynamics, from which trim solutions and simplified models can be determined. This model is of the highest fidelity, and could be used to explore perching

Table 5.3 Modal parameters of the decoupled linear time-periodic model

Eigenvalue	Magnitude	Eigenvector	Damping Ratio	Frequency [rad/s]
λ_i	$\|\lambda_i\|_2$	$\mathbf{v}_i$	ζ_i	ω_i
+1.04	1.04	$\begin{bmatrix} +0.13 \\ -0.17 \\ -0.78 \\ +0.59 \end{bmatrix}$	−1.00	0.21
+0.60	0.60	$\begin{bmatrix} -0.20 \\ -0.47 \\ +0.03 \\ +0.86 \end{bmatrix}$	+1.00	3.01
$+0.54 \pm 0.41j$	0.68	$\begin{bmatrix} -0.09 \mp 0.16j \\ +0.36 \mp 0.07j \\ -0.07 \mp 0.04j \\ +0.91 \end{bmatrix}$	+0.52	4.45
+1.04	1.04	$\begin{bmatrix} +0.59 \\ +0.19 \\ +0.14 \\ +0.77 \end{bmatrix}$	−1.00	0.23
+0.70	0.70	$\begin{bmatrix} +0.41 \\ -0.03 \\ -0.86 \\ +0.29 \end{bmatrix}$	+1.00	2.11
+1.84	1.84	$\begin{bmatrix} +0.28 \\ -0.42 \\ +0.24 \\ +0.83 \end{bmatrix}$	−1.00	3.60
+0.28	0.28	$\begin{bmatrix} -0.13 \\ -0.18 \\ +0.97 \\ -0.07 \end{bmatrix}$	+1.00	7.52

maneuvers and agile flight. Although it is difficult to design controllers for nonlinear systems, this model has been cast into a canonical form for Euler-Lagrange systems so that techniques such as passivity control can be applied in a straight forward manner [64, 62, 63, 66]. Additionally, techniques such as sliding mode or nonlinear damping could be used to attenuate the pitch oscillation, so that video

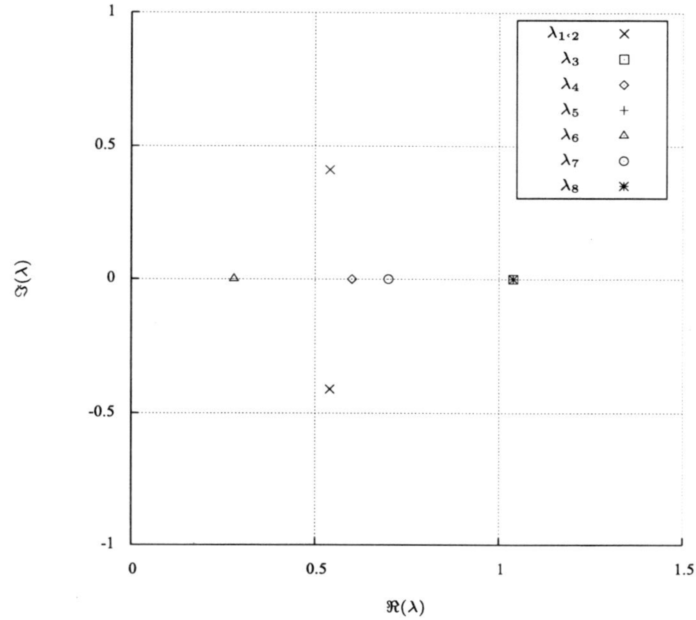

Figure 5.9 **Linear time-periodic model pole locations**

images are more stable and electronic components experience less vibration; similar control strategies have shown promise [27, 23, 36]. The heave motion, however, cannot be attenuated without morphing the wing structure to more gracefully generate the lift and thrust forces.

The LTP model describes the dynamics of perturbations about a trimmed straight and level mean flight trajectory. This model is still complex and must be determined and scheduled for each flight condition. However, controllers are simple enough for implementation on microprocessors, the dynamics still reflect the true motions, and analytical tools can be applied, e.g. the controllability result. This model could be used for an accurate model for navigation, or for inner-loop stabilization. Straight forward synthesis tools, such as a periodic Linear Quadratic Regulator (LQR) result, also exist [79]. For instance, solving LQR problem, using

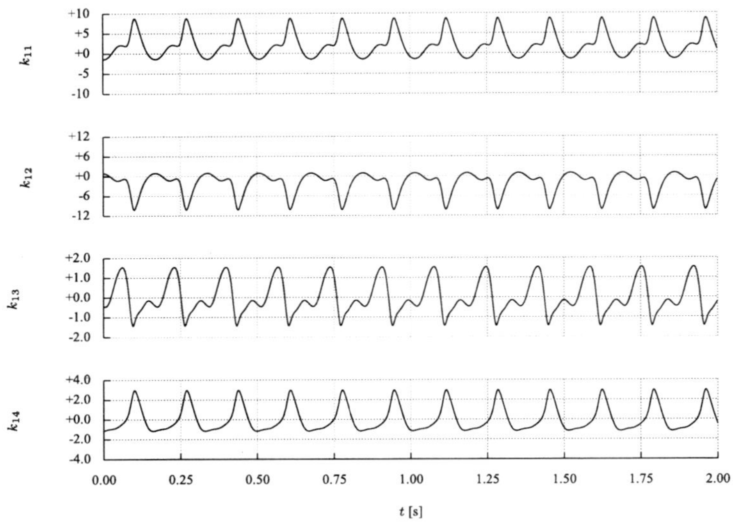

Figure 5.10 **Periodic control gains designed for the LTP system using LQR**

unity weighting matrices, results in the periodic controller gains shown in Figure 5.10. For an impulse applied at any point within the flapping cycle, the state responses settle within 0.6 seconds. Dietl and Garcia [51] have also shown that a discrete-time, infinite-horizon version of this control works well in simulation and suggests it is practical for flight control, despite control computations, controller gain memory requirements, and knowledge of the wing angle.

The simplest model presented is the LTI model, which describes the stroke-averaged dynamics of perturbations away from trimmed flight trajectories. This model enables a large set of analysis and synthesis tools. However, models must still be determined and scheduled for each flight condition, but more importantly these models average the significant effects of the wings flapping. For that reason, it should not be trusted for fast or accurate tasks, and should be relegated to slow, forgiving tasks such as navigation. Krashanitsa et al. were able to demonstrate navigation using a heuristically-tuned proportional/derivative control law [14]; however, the flight trajectories exhibited a large amount of variation between laps. Dietl and Garcia noted that although classical controllers are easy to implement, the system dynamics are periodically changing, so that these controllers are sometimes improving the flight performance, and sometimes degrading the flight performance. To control the pitch angle based on the identified LTI model, the proportional/integral/derivative controller

$$\delta_{lon}(t) = -6.67\Delta\theta(t) + 0.27\int_0^t \Delta\theta(\tau)d\tau - 2.03\Delta\dot{\theta}(t) \tag{5.22}$$

can be used, but only predicts a settling time of 32.7 seconds. Applying the LQR controller, again solved with unity weighting matrices,

$$\delta_{lon}(t) = -0.23\Delta\theta(t) + 0.14\Delta u(t) - 0.02\Delta w(t) - 0.15\Delta q(t) \tag{5.23}$$

can more effectively capture cross couplings and predicts settling times under three seconds. Another possibility is to use this simple model in conjunction with a robust control scheme, such as H_∞ or μ-synthesis, or rather a robust adaptive control law such as L_1, to learn how to compensate for the changing system dynamics [52].

5.5 Chapter summary

This chapter presented an investigation into the dynamics of the identified ornithopter model using simulation tools. The model was programmed into both FORTRAN and MATLAB simulation environments, which offered trade-offs between computational speed and the availability of analysis tools. A time step of 10^{-4} s was determined necessary for convergence with a fourth order Runge-Kutta routine.

The periodic forcing of the wings admits trimmed flight trajectories that are limit cycle oscillations in the state space.

Simulations from initial conditions suggest that these limit cycles are unstable, and a simple method was introduced for locating initial conditions and trimmed control settings that would put the ornithopter state on the limit cycle. This method was used to trim the ornithopter for straight and level mean flight for a variety of flapping frequencies. As the flapping frequency increased, the tail angle became more negative, the mean forward speed increased, the mean pitch angle and pitch oscillations decreased, the altitude variation decreased and the heave velocity oscillations increased. Additionally a bifurcation is evident around 7 Hz, due to a shape change in the pitch dynamics and heave dynamics limit cycles.

Linear time-periodic and linear time-invariant models were numerically determined from the nonlinear model. The LTI model exhibited a stable and unstable short period mode, while the LTP model showed a stable short period mode, a stable subsidence mode, and an unstable divergence mode.

Finally, implications for control strategies were discussed. For high precision or rapid maneuvering control, the nonlinear model should be used. Although not a straight forward process, the model has been cast into a canonical form for nonlinear control. The LTP model is suitable for accurate regulation of trimmed flight conditions, and the LTI model is only suitable for simple tasks such as heading or altitude-hold navigation.

6

Concluding remarks

DOI: 10.1533/9780857094674.95

Abstract: This chapter summarizes the work presented within this book.

Key Words: summary

6.1 Summary of work

Flight dynamics models are needed for flapping-wing aircraft, so that the dynamics can be better understood and flight controllers can be designed to perform mission scenarios ranging from robust outdoor navigation to indoor agile flight and perching. This work contributes a nonlinear model of the flight dynamics, as well as an investigation into linearizing the model and the ramifications on control design.

Chapter 2 presented characterizations of an ornithopter flight test platform. The aircraft design and operation are first discussed. Then an experiment is presented where the ornithopter was fitted with a custom avionics package to measure state variables, in real time, during trimmed straight and level mean flight. Afterwards, the variation in the mass distribution is examined using computer aided design software. An experiment is then summarized where aerodynamic and configuration data were measured in a constrained quasi-hover condition. The implications for flight dynamics modeling are then discussed.

The results of Chapter 2 suggest a nonlinear multibody model of the ornithopter vehicle dynamics. Chapter 3 begins with a review of the available tools for obtaining multibody models. Based on this review, a Lagrangian type model was selected and the derivation is presented using the Boltzmann-Hamel equations. The equations of motion are then cast into a canonical form for nonlinear control and the model is discussed.

Chapter 4 presents the system identification work completed to identify parts of the ornithopter flight dynamics model not known a priori. A review of system identification techniques and methods is presented. Afterwards, a wind tunnel test is discussed where lift and drag values were calculated for the ornithopter tail. Finally an experiment is presented using a visual tracking system to obtain flight test data with which an aerodynamic model for the wings was identified.

Chapter 5 presents flight simulation results obtained using the flight dynamics model developed in Chapters 3 and 4. A method for finding trim solutions is presented and was employed to trim the ornithopter model for straight and level mean flight. The model was then linearized about the trim trajectory, resulting in a canonical time-invariant model as well as a time-periodic model.

6.2 Summary of modeling assumptions

- Vehicle dynamics are approximated by three interconnected bodies
  The ornithopter multibody model consisted of a fuselage and two wings, which comprise 95.1% of the mass and captures the inertial effects of the flapping wings. Extra bodies for the tail surface were not modeled, as the tail has only 4.9% of the mass, moves slowly relative to the flapping wings, and does not create a significant change in the mass distribution. Instead, the tail mass and inertia was absorbed into the fuselage. Similarly, no additional benefit was observed from modeling the four-bar mechanism used to flap the wings.
- Structural dynamics in the vehicle dynamics are ignored

Structural deformations have been observed in the fuselage, wing spars, and wing sail membrane; however, these bodies are assumed to be rigid in order to develop models for control. Torsion and lateral bending modes in the fuselage are at higher frequencies than used in control. Additionally, tip displacements of the wing spars are relatively small, less than 8% of the wingspan [20]. Finally, although the aft section of the wings deflect to produce thrust, this portion contains less than 2.3% of the vehicle mass and does not significantly vary the mass distribution.

- Actuator dynamics are negligible
  It is assumed that the pilot or autopilot directly controls the flapping frequency and orientation of the tail. This bypasses the actuator dynamics, presented in Appendix A.2, and thus any transmission delays, time lags, backlash, mechanical slop, and current/torque saturation limits. These models can be easily incorporated to increase model fidelity. Momentum sources from the spinning motors and gears are assumed negligible.
- Tail aerodynamics are steady and do not interact with the wings
  The downwash from the wings onto the tail is assumed negligible due to the placement of the tail between and above the wings. Additionally, limited actuator and pilot bandwidths led to steady models of the tail aerodynamics. The full effect of downwash from the wings, unsteady aerodynamic effects from the body pitching motion, and flexibility in mechanisms is not currently known.

6.3 Summary of original contributions

- Flight data suitable for modeling
  This work presented flight data of an ornithopter, sampled at a bandwidth that illuminated the effects of the flapping wings. Trimmed flight data showed pitch oscillations up to 4.97 rad/s and heave accelerations up to 41.7 m/s^2, which

require nonlinear multibody models to capture. Additionally, harmonic responses were seen in the data, indicating either structural modes or nonlinearities. High accelerations preclude the use of traditional attitude determination methods and require model-based techniques.

- Modeling of mass distribution variation due to wings flapping
  The effects of the flapping motion on the mass distribution of the ornithopter was investigated using a CAD model. The center of mass travels 10.3 cm in the vertical direction, the moments of inertia can vary up to 53.6%, and the inertia rates are significant. Multibody models are required to capture these variations.
- Derivation of rigid multibody vehicle dynamics
  It was determined that a three-body model was sufficient to model the ornithopter. Vehicle dynamics were derived using the Boltzmann-Hamel equations, and were cast into a canonical form used for the nonlinear control of Euler-Lagrange systems.
- Tail aerodynamics system identification
  Wind tunnel tests were conducted on the ornithopter tail, having free stream velocities, angles of attack, and angles of sideslip ranging between 4.50 to 6.33 m/s, −0.88 to −0.35 rad, and ±0.57 rad, respectfully. System identification methods were applied to determine a model for the aerodynamic coefficients.
- Flight testing and wing aerodynamics system identification
  The ornithopter flight data, obtained using a visual tracking system, was presented. This data and the vehicle dynamics model were used to explore variations in angle of attack, Reynolds number, reduced frequency, velocity, and structural deformations over the wing and throughout the wing stroke cycle. Additionally, force and moment sources were identified throughout the wing stroke. Finally, this data was used to determine a wing aerodynamics model obtained from flight data using system identification methods.
- Models for stability and control

In addition to the full nonlinear model, a linear time-invariant model and a linear time-periodic model was presented, with modeling implications and ramifications for control.

6.4 Recommendations for future research

- Aerodynamic model expansion
  The aerodynamic model presented in Chapter 4 is valid for a relatively small region of the flight envelope. The tail was tested in steady conditions for flow velocities between 4.50 and 6.33 m/s, angles of attack between −0.88 and −0.35 rad, and angles of sideslip between ±0.57 rad. The model could be improved by increasing these ranges and be looking at the dynamic effects of actuating the tail.
  The flight test was conducted for a flapping frequency of 5.91 Hz forward speeds between 9 and 10 m/s. For these tests the tail was fixed, and so the only dynamics excited were a result of the flapping wings. More testing should be performed where the lateral dynamics are excited, the throttle changed, and the tail excited, in order to expand the aerodynamic model of the wings. Additionally, longer sets of data are needed to excite the forward speed dynamics in order to model the thrust force more accurately. In order to complete these tasks, a larger flight space is needed. This could be accomplished using a large building instrumented with numerous tracking cameras, or perhaps by flying outdoors with more instrumentation; however, problems still arise when using air data booms and trying to solve the attitude problem due to the pitching and heaving motion.
- Wing/tail aerodynamic interactions
  In this work, it was assumed that the wing and tail aerodynamics do not interact. Although the tail is centered between and located above the wings, it is currently unknown how much downwash from the wing impinges

on the tail. Flow visualization or particle image velocimetry experiments would provide this information, which would then be used to update the aerodynamic model structure of the tail. Additionally, the flight test mentioned above where the wings and tail were both excited could also provide this insight.

- Aeroelastic effects
  It is known that aeroelastic effects play a role in the flight dynamics of ornithopters; however, this work focused on models for flight control design and neglected the aeroelastic effects. The nonlinear model fidelity could be improved by incorporating finite element models to capture the fluid/structure interactions, as in Lee et al. [28], so that the effect of aeroelasticity could be assessed.
- Control law design
  Control laws can now be designed from the models presented in this work. It remains then to implement these in flight controllers and evaluate their performance. This capability would allow direct comparison of controllers developed from the LTI and LTP models, and would facilitate the development of perching and aggressive maneuvering experiments.
- Integrated platform for morphing
  Provided a flight dynamics model and a stabilized flight platform, work on morphing wing structures could progress. Morphing would allow more degrees of freedom, with which the ornithopter could perform more efficient and agile flight.

Appendix A: Field calibration of inertial measurement units

Abstract: Inertial measurement units (IMUs) consist of orthogonal triads of magnetometers, gyroscopes, and accelerometers, and are used in autopilots to provide information on the vehicle pose and motion. The IMUs used on unmanned aircraft suffer from the low signal-to-noise ratios and strong temperature dependencies typical of Microelectromechcanical systems (MEMS) devices, requiring tedious laboratory testing or on-board state estimators for calibration. In this appendix, a method is presented for estimating IMU calibration constants, based on three successive least-squares computations. Memory and computation requirements are low enough for implementation on 8 bit microprocessors, and experimental data shows adequate calibration results.

Key Words: IMU calibration, least-squares, flight data

A.1 Theory and method

The IMU provides measurements of the magnetic field $\mathbf{h}_d^0$, linear acceleration a_d^0, and rotational velocity ω_d^0, corrected to the aircraft body frame [55, 37]. The subscript $_d$ is used to denote measurements in raw digital format, e.g. an integer between 0 and 65535. Assuming a constant temperature, as unmanned aircraft have limited flight durations and altitude ceilings, the calibration model is

$$\mathbf{y}^0 = \mathbf{\Lambda} \mathbf{y}_d^0 + \mathbf{b} + \mathbf{n} \tag{6.1}$$

where $\mathbf{y}$ is the measurement, $\Lambda = diag\{\lambda_x, \lambda_y, \lambda_z\}$ is a matrix of scale factors, $\mathbf{b}$ is a vector of biases, and $\mathbf{n}$ is a vector of additive measurement noise.

A block diagram of the calibration process is shown in Figure A.1. Data is collected from a maneuver where quick angular deviations are made to excite each sensor channel without introducing large translational accelerations. Existing least-square calibration algorithms can be used to calibrate the magnetometer and accelerometer [80, 81], which can then estimate the vehicle attitude [58, 82].

Substituting the sensor model (6.1), the rotational kinematic equation of motion is

$$\dot{\boldsymbol{\eta}}^I = \mathbf{\Phi} \left(\mathbf{\Lambda}_\omega \boldsymbol{\omega}_d^0 + \mathbf{b}_\omega + \mathbf{n}_\omega \right) \tag{6.2}$$

which is a least-squares problem that solves for the gyroscope calibration parameters using the attitude estimates in a certainty-equivalence fashion. Using a quaternion to represent attitude, (6.2) can be solved quickly and accurately by using only the scalar portion of the quaternion.

A.2 Results

The avionics package shown in Figure 2.4, was used to record measurements from an IMU, through a 16 bit analog to digital converter, at 185 Hz. A visual tracking system was employed to obtain high accuracy estimates of the avionics orientation, derived from spatial measurements of ten retro-reflective markers placed on the avionics board. Hardware specifications for the IMU and the visual tracking system are given in Table A.1. The maneuver chosen for analysis was a

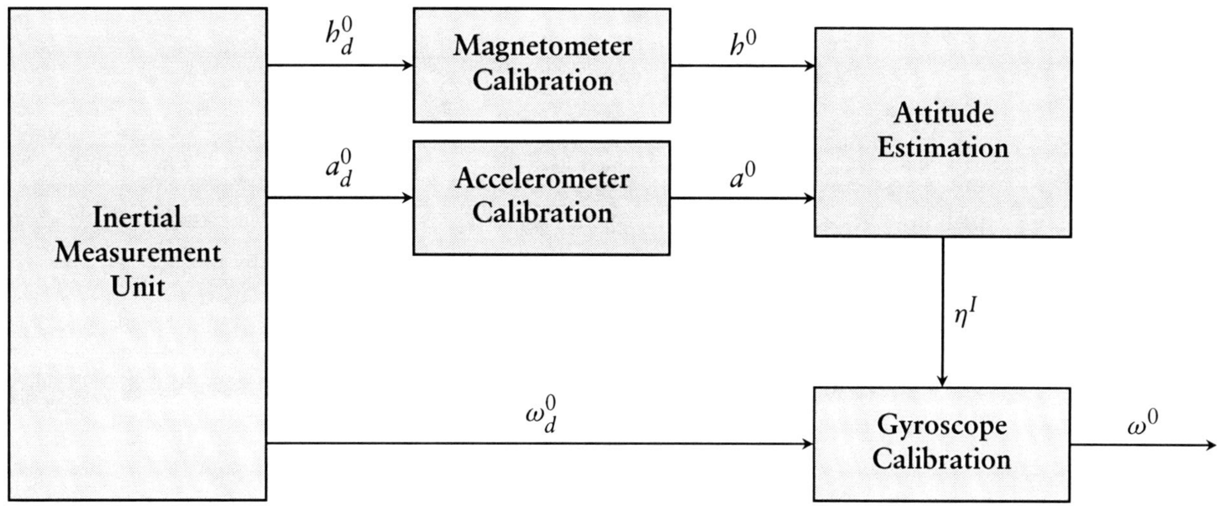

Figure A.1 IMU calibration method block diagram

Table A.1 Measurement specifications for avionics and visual positioning system

Sensor	Symbol	Range	Resolution	Error	Units
time	t	-	25.6×10^{-6}	-	s
magnetometer	$\mathbf{h}^0$	± 3.17	48.3×10^{-6}	5.83×10^{-3}	-
gyroscope	ω^0	± 5.24	0.16×10^{-3}	$0.14 \times 10^{+0}$	rad/s
accelerometer	a^0	± 98.1	3.88×10^{-3}	$0.05 \times 10^{+0}$	m/s^2
attitude	η^I	-	-	0.03×10^{-3}	rad

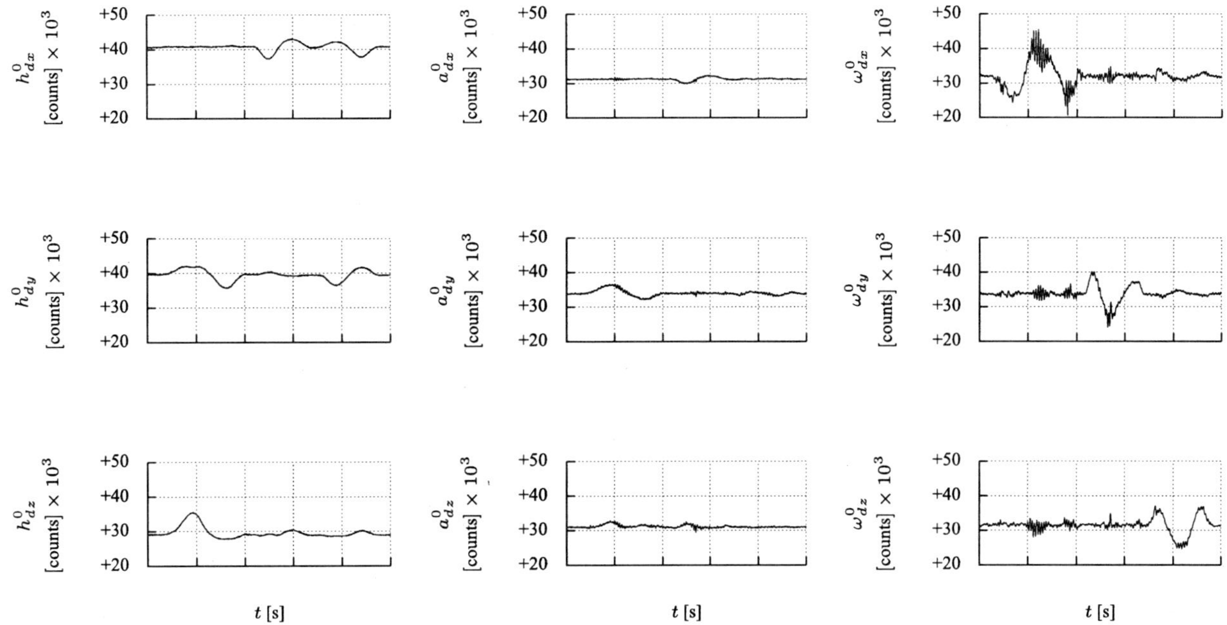

(a) magnetometer (b) accelerometer (c) gyroscope

Figure A.2 **Uncalibrated 16 bit IMU measurements from a roll-pitch-yaw maneuver**

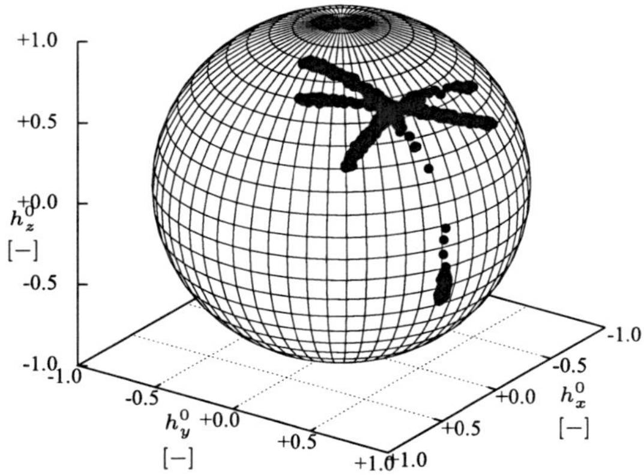

(a) magnetometer

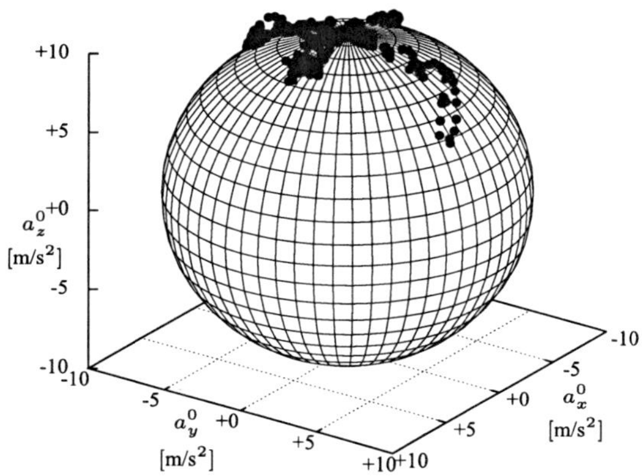

(b) accelerometer

Figure A.3 **Calibrated magnetometer and accelerometer signals with centered spheres**

sequence of rotational doublets along each sensing axes, as shown by the digitized measurements in Figure A.2.

The least-squares routine developed by Gebre-Egziabher [80] was used to calibrate the magnetometers and accelerometers. Field domain calibrations are shown in Figure A.3. Vehicle attitude was then computed using the QUEST algorithm, and equation-error was used to match the quaternion,

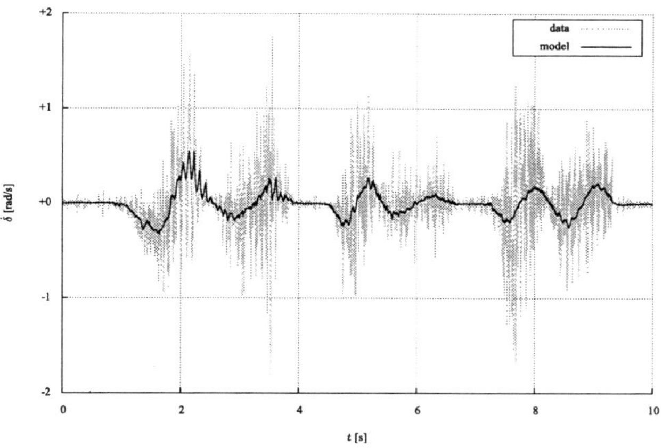

Figure A.4 **Model fit to the attitude kinematic equation using equation-error**

as shown in Figure A.4. Estimated parameters and standard errors are given in Table A.2. Magnetometer scale factors had a high error bound, perhaps due to the small amount of data used, but otherwise all estimated parameters had less than a 10% predicted error, indicating an accurate set of measurements.

The method presented in this section is a batch method, executed after the data has been collected. Alternatively, the least-squares routines could be implemented in a recursive fashion, as in [37, 83], so that the calibration constants and their associated error bounds evolve in real-time as more data becomes available. Estimating parameters in real-time would also allow for variations in the calibration due to temperature or flight condition changes over long periods of time.

Table A.2 **IMU calibration estimates and standard errors**

Parameter	Equation-Error	Multiplier
ϕ	$\hat{\phi} \pm \sigma(\hat{\phi})$	
λ_{hx}	$+0.16 \pm 2.05$	10^{-3}
λ_{hy}	$+0.14 \pm 1.89$	10^{-3}
λ_{hz}	$+0.16 \pm 2.04$	10^{-3}
b_{hx}	-6.36 ± 0.40	10^{+0}
b_{hy}	-4.98 ± 0.36	10^{+0}
b_{hz}	$+5.47 \pm 0.37$	10^{+0}
λ_{vx}	$+0.50 \pm 0.02$	10^{+0}
λ_{vy}	$+0.34 \pm 0.01$	10^{+0}
λ_{vz}	$+0.37 \pm 0.02$	10^{+0}
b_{vx}	-15.6 ± 3.17	10^{+0}
b_{vy}	-11.7 ± 2.74	10^{+0}
b_{vz}	-12.3 ± 2.81	10^{+0}
$\lambda_{\omega x}$	$+0.27 \pm 0.01$	10^{-3}
$\lambda_{\omega y}$	$+0.37 \pm 0.04$	10^{-3}
$\lambda_{\omega z}$	$+0.33 \pm 0.02$	10^{-3}
$b_{\omega x}$	$+8.58 \pm 0.46$	10^{+0}
$b_{\omega y}$	-12.5 ± 1.20	10^{+0}
$b_{\omega z}$	-10.2 ± 0.76	10^{+0}

Appendix B: Actuator dynamics system identification

Abstract: The ornithopter has on board a Feigao GF2030 brushless motor to drive the flapping motion, and two Hitec HS-56 servo motors to orient the tail. Due to time lags observed in flight, models for the actuator dynamics were identified to determine actuator bandwidths and to enable higher fidelity models of the flight dynamics. This section presents the system identification experiments and modeling results.

Key Words: actuators, dynamic models, system identification

B.1 Experimental setup

The actuators were mounted to a reaction torque sensor for measuring the generated torque. The DC motor was fitted with a shaft encoder for a rotary speed measurement, while the servo motor internal potentiometer was measured for a rotary position measurement. A PIC18F452 microprocessor was used to generate input waveforms and to measure data at

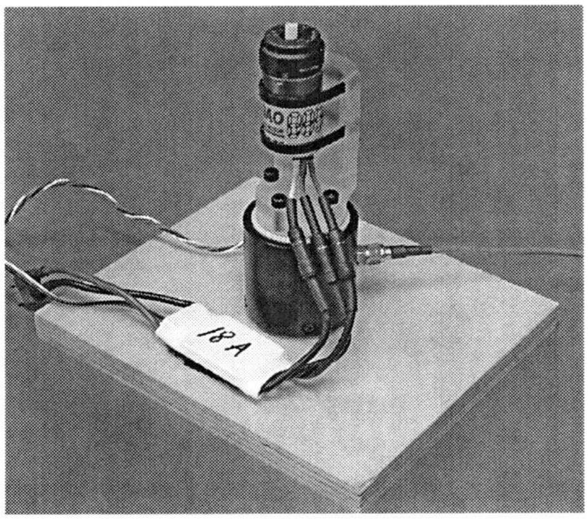

(a) DC motor

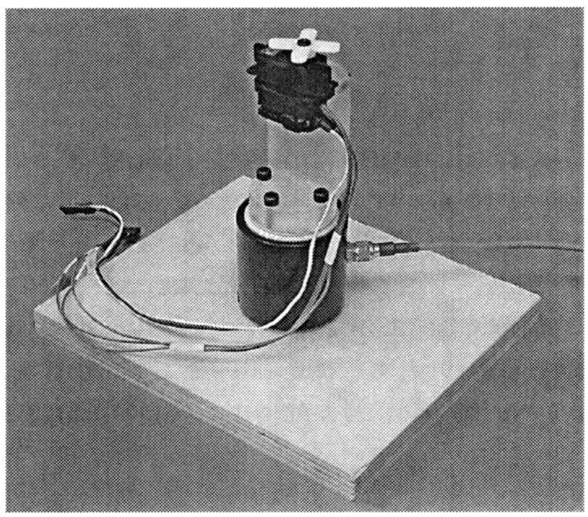

(b) servo motor

Figure B.1 **Actuators and instrumented test stand**

500 Hz using 16 bit analog to digital converters. The experimental setup is shown in Figure B.1 and measurement specifications are provided in Table B.1.

The data used for system identification are shown in Figure B.2. Both actuators were excited with a digital waveform that had an increasing and decreasing frequency.

Table B.1 **Actuator system identification measurement specifications**

Measurement	Range	Resolution	Error	Unit
time	-	0.05×10^{-3}	-	s
throttle input	0.00 to 1.00	0.25×10^{-3}	0.07×10^{-3}	-
angle input	$\pm 0.70 \times 10^{+0}$	0.35×10^{-3}	0.16×10^{-3}	rad
velocity	$\pm 3.77 \times 10^{+3}$	8.47×10^{-3}	$0.03 \times 10^{+0}$	rad/s
angle	$\pm 0.70 \times 10^{+0}$	0.16×10^{-3}	$0.13 \times 10^{+0}$	rad
torque	$\pm 0.35 \times 10^{+0}$	0.77×10^{-6}	0.90×10^{-3}	N·m

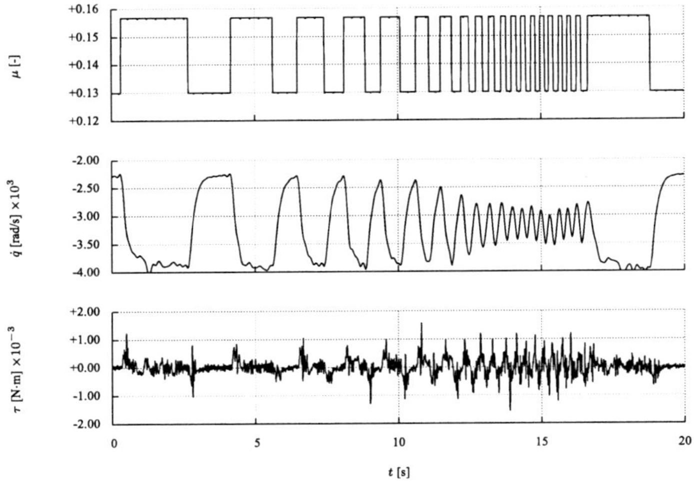

(a) DC motor

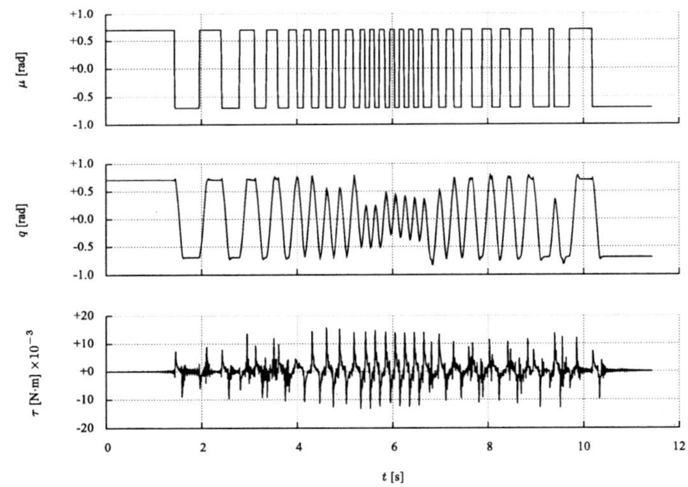

(b) servo motor

Figure B.2 **Measurements used for actuator system identification**

B.2 Results

DC and servo motors are typically linear transducers. Stepwise regression in the time domain using equation-error showed that a second order system

$$m\ddot{q} + c\dot{q} + kq + \tau = b\mu \tag{6.3}$$

was sufficient to model the systems. The mass encompasses the shaft inertia; damping describes effects of friction, back electro-mechanical force, current/torque coupling, and electrical resistance; stiffness models the position feedback used in the servo motor; τ is the shaft torque; and μ is the pilot stick input. Higher fidelity models such as Coulomb friction and third-order dynamics [84, 85, 62] made no significant increase in model fidelity.

Equation-error and output-error parameter identification methods were applied in both the time and frequency domains, as shown in Figure B.3. Estimated parameters are provided in Table B.2 and Table B.3. All models had coefficients of determination of 0.70 or above, and output-error parameter estimates were within statistical agreement. The DC motor has a bandwidth of 0.48 Hz and the servo motor 3.20 Hz.

B.3 Coupling to vehicle dynamics

The rigid body dynamic equations (3.18) can be written as

$$M_b\dot{v} + C_b v + g + a = \boldsymbol{\tau}. \tag{6.4}$$

The actuator dynamics (6.3) can be written in matrix form, in terms of the ornithopter state variables as

$$M_a\dot{v} + C_a v + G_a\boldsymbol{\tau} + K_a p = B_a\boldsymbol{\mu} \tag{6.5}$$

where $\mathbf{G_a}$ is a matrix of gear ratios, relating the actuator outputs to the ornithopter joint angles. Substituting (6.4) into (6.5) and rearranging yields

$$(M_a + G_aM_b)\dot{v} + (C_a + G_aC_b)v + (G_ag + G_aa + K_ap) = B_a\boldsymbol{\mu} \tag{6.6}$$

which is the same canonical form as the equations of motion described in Chapter 3.

Table B.2 **DC motor parameter estimates and standard errors**

Method	Mass	Damping	Stiffness	Fit
	$m \times 10^{-3}$	$c \times 10^{-3}$	$k \times 10^{+0}$	R^2
time equation-error	$+0.28 \pm 0.01$	$+0.92 \pm 0.06$	-64.1 ± 3.08	0.69
frequency equation-error	$+0.23 \pm 0.01$	$+0.89 \pm 0.02$	-58.4 ± 1.30	0.78
time output-error	$+0.22 \pm 0.01$	$+1.01 \pm 0.03$	-64.6 ± 2.07	0.98
frequency output-error	$+0.20 \pm 0.01$	$+0.92 \pm 0.04$	-58.0 ± 2.98	0.99

Table B.3 Servo motor parameter estimates and standard errors

Method	Mass	Damping	Stiffness	Fit
	$m \times 10^{-3}$	$c \times 10^{-3}$	$k \times 10^{+0}$	R^2
time equation-error	$+0.01 \pm 0.00$	$+0.15 \pm 0.01$	$+2.22 \pm 0.27$	0.81
frequency equation-error	$+0.03 \pm 0.00$	$+0.15 \pm 0.01$	$+2.15 \pm 0.03$	0.96
time output-error	$+0.04 \pm 0.00$	$+0.53 \pm 0.04$	$+7.12 \pm 0.82$	0.98
frequency output-error	$+0.04 \pm 0.00$	$+0.57 \pm 0.01$	$+5.58 \pm 0.14$	0.97

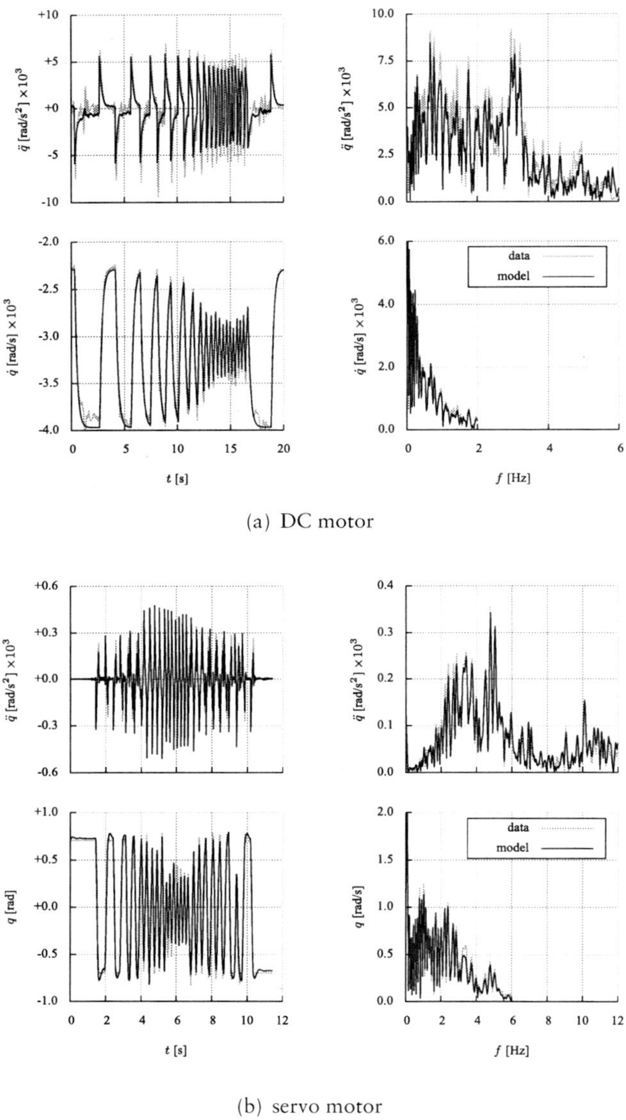

Figure B.3 **Model fits for actuator dynamics using equation-error and output-error in the time and frequency domains**

Appendix C: Equations of motion for single-body flight vehicles

Abstract: Many flight vehicles, including conventional fixed-wing aircraft, rotary-wing aircraft, and spacecraft, can often be modeled as a single body. Derivation of the equations of motion is most commonly performed using the Newton-Euler approach [1, 37]; however, to provide familiarity with the energy approach presented in Chapter 3, the equations of motion for a single-body vehicle are derived here using the Boltzmann-Hamel equations.

Key Words: rigid body dynamics, Boltzmann-Hamel equations, conventional aircraft

Consider the single-body vehicle in Figure 2.1, having position and velocity state variables

$$\mathrm{p} = \begin{bmatrix} \mathbf{r}_{0,I}^{I} \\ \boldsymbol{\eta}_{0,I}^{I} \end{bmatrix} \qquad \mathrm{v} = \begin{bmatrix} \boldsymbol{\nu}_{0,I}^{0} \\ \boldsymbol{\omega}_{0,I}^{0} \end{bmatrix}. \tag{6.7}$$

Written in terms of these state variables, the kinetic energy

$$T(\mathrm{p}, \mathrm{v}) = \frac{1}{2}(\boldsymbol{\nu}_{0,I}^{0})^{T}(m_0\mathbb{I})(\boldsymbol{\nu}_{0,I}^{0}) + \frac{1}{2}(\boldsymbol{\omega}_{0,I}^{0})^{T}\mathbf{I}_{0}^{0}(\boldsymbol{\omega}_{0,I}^{0}) \tag{6.8}$$

is the summation of the system energy. Using algebraic manipulations, (6.8) can be rearranged into the quadratic form

$$T(\mathbf{p}, \mathbf{v}) = \frac{1}{2}\mathbf{v}^T\mathbf{M}\mathbf{v} \tag{6.9}$$

to admit the generalized mass matrix

$$\mathbf{M} = \begin{bmatrix} m_0\mathbb{I} & \mathbf{0} \\ \mathbf{0} & \mathbf{I}_0^0 \end{bmatrix}. \tag{6.10}$$

As the mass matrix does not depend on the position states of the aircraft, the coupling matrix given by (3.22) reduces to

$$\boldsymbol{C(p, v)} = \dot{\boldsymbol{M}} + \sum_{k=1}^{n_v} T v_k \boldsymbol{\Gamma}_k \tag{6.11}$$

where the number of velocity states n_v is six. Partial differentiation of (6.9) with respect to the velocity states reveals that the derivative term in (6.11) refers to k^{th} entry of the $(n_v \times 1)$ row vector $\mathbf{v}^T\mathbf{M}$. The second term in (6.11) is the set of Hamel coefficient matrices Γ_k, which can be computed using (3.8), (3.14), and (3.15) as

$$\boldsymbol{\Gamma}_1 = \begin{bmatrix} 0 & 0 & 0 & 0 & 0 & 0 \\ 0 & 0 & 0 & 0 & 0 & 1 \\ 0 & 0 & 0 & 0 & -1 & 0 \\ 0 & 0 & 0 & 0 & 0 & 0 \\ 0 & 0 & 1 & 0 & 0 & 0 \\ 0 & -1 & 0 & 0 & 0 & 0 \end{bmatrix} \quad \boldsymbol{\Gamma}_2 = \begin{bmatrix} 0 & 0 & 0 & 0 & 0 & -1 \\ 0 & 0 & 0 & 0 & 0 & 0 \\ 0 & 0 & 0 & 1 & 0 & 0 \\ 0 & 0 & -1 & 0 & 0 & 0 \\ 0 & 0 & 0 & 0 & 0 & 0 \\ 1 & 0 & 0 & 0 & 0 & 0 \end{bmatrix}$$

$$\boldsymbol{\Gamma}_3 = \begin{bmatrix} 0 & 0 & 0 & 0 & 0 & 0 \\ 0 & 0 & 0 & 0 & 0 & 1 \\ 0 & 0 & 0 & 0 & -1 & 0 \\ 0 & 0 & 0 & 0 & 0 & 0 \\ 0 & 0 & 1 & 0 & 0 & 0 \\ 0 & -1 & 0 & 0 & 0 & 0 \end{bmatrix} \quad \boldsymbol{\Gamma}_4 = \begin{bmatrix} 0 & 0 & 0 & 0 & 0 & -1 \\ 0 & 0 & 0 & 0 & 0 & 0 \\ 0 & 0 & 0 & 1 & 0 & 0 \\ 0 & 0 & -1 & 0 & 0 & 0 \\ 0 & 0 & 0 & 0 & 0 & 0 \\ 1 & 0 & 0 & 0 & 0 & 0 \end{bmatrix}$$

$$\boldsymbol{\Gamma}_5 = \begin{bmatrix} 0 & 0 & 0 & 0 & 0 & 0 \\ 0 & 0 & 0 & 0 & 0 & 1 \\ 0 & 0 & 0 & 0 & -1 & 0 \\ 0 & 0 & 0 & 0 & 0 & 0 \\ 0 & 0 & 1 & 0 & 0 & 0 \\ 0 & -1 & 0 & 0 & 0 & 0 \end{bmatrix} \quad \boldsymbol{\Gamma}_6 = \begin{bmatrix} 0 & 0 & 0 & 0 & 0 & -1 \\ 0 & 0 & 0 & 0 & 0 & 0 \\ 0 & 0 & 0 & 1 & 0 & 0 \\ 0 & 0 & -1 & 0 & 0 & 0 \\ 0 & 0 & 0 & 0 & 0 & 0 \\ 1 & 0 & 0 & 0 & 0 & 0 \end{bmatrix} \tag{6.12}$$

regardless of the attitude parametrization. Applying these terms yields the dynamic coupling matrix

$$\mathbf{C}(\mathbf{p},\mathbf{v}) = \begin{bmatrix} \dot{m}_0 & \\ & \dot{\boldsymbol{I}}_0^0 \end{bmatrix} + \begin{bmatrix} 0 & -m_0 S(\boldsymbol{\nu}_{0,I}^0) \\ -m_0 S(\boldsymbol{\nu}_{0,I}^0) & -S(\mathbf{I}_0^0 \boldsymbol{\omega}_{0,I}^0) \end{bmatrix} \tag{6.13}$$

for a single body vehicle. Furthermore, if the vehicle can be assumed rigid so that it neither changes mass nor inertia, the mass rate matrix in (6.11) and (6.13) is null.

The remaining forces on the vehicle are those imparted by the environment through which it passes. The effects of gravity can be modeled by first writing the potential energy of the aircraft

$$U(\mathbf{p}) = -m_0 (\mathbf{r}_{0,I}^I)^T \boldsymbol{g}^I \tag{6.14}$$

and then using (3.26) to form the generalized gravitational forces vector

$$\mathbf{g}(\mathbf{p}) = \begin{bmatrix} m_0 \mathbf{R}^{0,I} \boldsymbol{g}^I \\ 0 \end{bmatrix}. \tag{6.15}$$

Combining terms and adding aerodynamic contributions $\mathbf{a}(\mathbf{p},\mathbf{v})$, the full nonlinear dynamics of the rigid body are

$$M(p)\dot{\boldsymbol{v}} + C(p,v)v + g(p) + a(p,v) = \boldsymbol{\tau} \tag{6.16}$$

where τ is a vector of exogenous forces and torques. This parametrization of the dynamics satisfies the property used in passivity control design that $\dot{\boldsymbol{M}} - 2C$ is a skew-symmetric matrix [66, 59, 64].

Appendix D: Linearization of a conventional aircraft model

Abstract: This section presents the linearization of a conventional aircraft model to provide a baseline for comparison with the results presented in Chapter 5. The aircraft chosen was the single-seat fighter F-16 Fighting Falcon, which has a nominal 9279 kg mass and 9.96 m wing span. A nonlinear dynamics model is contained within SIDPAC, and is documented in references [37, 2, 86].

Key Words: linearization, central finite differences, conventional aircraft

The aircraft was set at a 3048 m altitude and a 0.09 rad pitch angle. A modified Newton-Raphson algorithm was then used to solve (5.2) for the remaining trim states, which were a 137 m/s airspeed, a 19.5% throttle setting, and a −0.08 rad elevator deflection. The center of mass is at the mean aerodynamic quarter-chord location to synthetically increase open loop stability.

Numerical linearization using finite differences resulted in the time-invariant model where δ_e is the elevator deflection, δ_a is the aileron deflection, and δ_r is the rudder deflection.

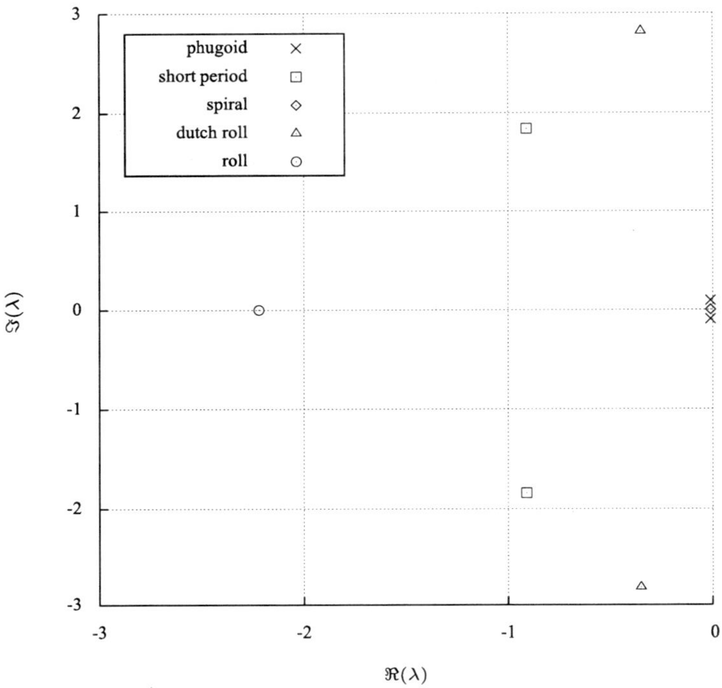

Figure D.1 **Linearized F-16 model pole locations**

In this flight condition, the block-diagonal structure of the system matrices results in a decoupling of the longitudinal and lateral dynamics. Kinematic integrations are seen in the equations for the pitch angle and bank angle. The local acceleration due to gravity, 9.77 m/s^2, appears in the equations for longitudinal and lateral velocity. The elevator primarily affects the pitch rate, but also the longitudinal and heave velocities. The aileron and rudder primarily affect the roll and yaw rate, respectively, but also couple into the other lateral states.

Poles are shown in Figure D.1, and modal parameters of the decoupled subsystems are provided in Table D.1. The longitudinal dynamics have two oscillatory, under-damped

Table D.1 **Modal parameters of the decoupled F-16 linear model**

Mode	Eigenvalue λ_i	Eigenvector $\mathbf{v}_i$	Damping Ratio ζ_i	Frequency [rad/s] ω_i
phugoid	$-0.01 \pm 0.09j$	$\begin{bmatrix} -0.00 \mp 0.01j \\ +0.99 \\ +0.05 \mp 0.00j \\ +0.00 \pm 0.00j \end{bmatrix}$	+0.11	0.09
short period	$-0.91 \pm 1.84j$	$\begin{bmatrix} +0.01 \mp 0.00j \\ -0.07 \pm 0.03j \\ +0.99 \\ -0.00 \pm 0.01j \end{bmatrix}$	+0.44	2.06
spiral	-0.01	$\begin{bmatrix} +0.89 \\ +0.44 \\ -0.02 \\ +0.06 \end{bmatrix}$	+1.00	0.01
dutch roll	-0.35 ± 2.82	$\begin{bmatrix} +0.01 \pm 0.01j \\ +0.99 \\ -0.03 \pm 0.04j \\ -0.00 \mp 0.02j \end{bmatrix}$	+0.12	2.84
roll	-2.22	$\begin{bmatrix} -0.34 \\ -0.56 \\ +0.75 \\ +0.03 \end{bmatrix}$	+1.00	2.22

modes. The phugoid mode is slower and involves primarily the forward velocity, whereas the short period mode is faster and involves primarily the heave velocity. The lateral dynamics comprise three modes. The spiral mode is slow, first-order mode that involves primarily the bank angle and side velocity. The roll mode is fast, first-order mode that involves mostly the roll rate. The dutch roll mode is a fast, second-order mode using mostly the side velocity.

$$
\begin{pmatrix} \Delta\dot{\theta} \\ \Delta\dot{u} \\ \Delta\dot{w} \\ \Delta\dot{q} \\ \Delta\dot{\phi} \\ \Delta\dot{v} \\ \Delta\dot{p} \\ \Delta\dot{r} \end{pmatrix}
=
\begin{pmatrix}
ccc| +0.00 & +0.00 & +0.00 & +1.00 & +0.00 & +0.00 & +0.00 & +0.00 \\
-9.77 & -0.01 & +0.06 & -11.5 & +0.00 & +0.00 & +0.00 & +0.00 \\
+0.00 & -0.08 & -0.68 & +127. & +0.00 & -0.00 & +0.00 & +0.00 \\
+0.00 & +0.00 & -0.03 & -1.15 & +0.00 & +0.00 & +0.00 & +0.00 \\
+0.00 & +0.00 & +0.00 & -0.00 & +0.00 & +0.00 & +1.00 & +0.00 \\
+0.00 & -0.00 & -0.00 & +0.00 & +9.77 & -0.21 & +12.1 & -136. \\
+0.00 & -0.00 & +0.00 & -0.00 & +0.00 & -0.16 & -2.37 & +0.60 \\
+0.00 & -0.00 & +0.00 & -0.00 & +0.00 & +0.05 & -0.05 & -0.35
\end{pmatrix}
\begin{pmatrix} \Delta\theta \\ \Delta u \\ \Delta w \\ \Delta q \\ \Delta\phi \\ \Delta v \\ \Delta p \\ \Delta r \end{pmatrix}
+
\begin{pmatrix}
| +0.00 & +0.00 & +0.00 \\
+2.08 & +0.00 & +0.00 \\
-11.2 & +0.00 & +0.00 \\
-6.43 & +0.00 & +0.00 \\
+0.00 & +0.00 & +0.00 \\
+0.00 & +1.54 & +4.20 \\
+0.00 & -25.3 & +4.28 \\
+0.00 & -1.11 & -2.28
\end{pmatrix}
\begin{pmatrix} \Delta\delta_e \\ \Delta\delta_a \\ \Delta\delta_r \end{pmatrix}
\tag{6.17}
$$

References

1. D. McRuer, I. Ashkenas, and D. Graham, *Aircraft Dynamics and Automatic Control*, Princeton University Press, 1973.
2. B. Stevens and F. Lewis, *Aircraft Control and Simulation*, Wiley, 2003.
3. B. Etkin and L. Reid, *Dynamics of Flight: Stability and Control*, Wiley, 1996.
4. T. Mueller and J. DeLaurier, *An Overview of Micro Air Vehicle Aerodynamics*, vol. 195 of *Progress in Aeronautics and Astronautics*, chapter 1, pp. 1–10, AIAA, 2001.
5. F. Bohorquez, D. Pines, and P. Samuel, "Small rotor design optimization using blade element momentum theory and hover tests", *Journal of Aircraft*, vol. 47, no. 1, pp. 268–283, January–February 2010.
6. U. Pesavento and Z. Wang, "Flapping wing flight can save aerodynamic power compared to steady flight", *Physical Review Letters*, vol. 103, no. 11, pp. 1–4, September 2009.
7. R. Malhan, M. Benedict, and I. Chopra, "Experimental investigation of a flapping wing concept in hoverand forward flight for micro air vehicle applications", Phoenix, AZ, May 2010, American Helicopter Society, 66th Annual Forum.
8. P. Seshadri, M. Benedict, and I. Chopra, "Experimental investigation of an insect-based flapping wing hovering

micro air vehicle", San Francisco, CA, January 2010, American Helicopter Society, Aeromechanics Specialists' Conference.

9. T. Liu, K. Kuykendoll, R. Rhew, and S. Jones, "Avian wings", Portland, OR, June 2004, AIAA, number 2004–2186 in Aerodynamic Measurement Technology and Ground Testing Conference.
10. J. Videler, *Avian Flight*, Oxford Ornithology Series. Oxford University Press, 2005.
11. D. Alexander, *Nature's Flyers: Birds, Insects, and the Biomechanics of Flight*, The Johns Hopkins University Press, 2002.
12. J. Delaurier, "An ornithopter wing design", *Canadian Aeronautics and Space Journal*, vol. 40, no. 1, pp. 10–18, March 1994.
13. H. Räbinger, "Die Entwicklung der EV-Schlagflügelmodelle", Tech. Rep., Nürnberg, Germany, 2002.
14. R. Krashanitsa, D. Silin, S. Shkarayev, and G. Abate, "Flight dynamics of a flapping-wing air vehicle", *International Journal of Micro Air Vehicles*, vol. 1, no. 1, pp. 35–49, 2009.
15. M. Keennon and J. Grasmeyer, "Development of the black widow and microbat mavs and a vision for the future of mav design", Dayton, OH, July 2003, AIAA, number 2003–2901 in International Air and Space Symposium and Exposition conference.
16. Z. Jackowski, "Design and construction of an autonomous ornithopter", Master's thesis, Massachusetts Institute of Technology, 2009.
17. E. Avadikian and J. Hubbard, "Ornithopter development and optimization guide", Tech. Rep., University of Maryland, 2010.
18. J. Leishman, *Principles of Helicopter Aerodynamics*, Cambridge University Press, 2nd edition, 2006.
19. R. Harmon, J. Grauer, J. Hubbard, J. Conroy, S. Humbert, J. Sitaraman, and B. Roget, "Experimental determination of ornithopter membrane wing shapes used for

simple aerodynamic modeling", AIAA, 2008, number 2008–6237 in Applied Aerodynamics Conference.

20. R. Harmon, "Aerodynamic modeling of a flapping membrane wing using motion tracking experiments", Master's thesis, University of Maryland, 2008.
21. T. Rashid, "The flight dynamics of a full-scale ornithopter", Master's thesis, University of Toronto, 1995.
22. C. Orlowski, A. Girard, and W. Shyy, "Open loop pitch control of a flapping wing micro-air vehicle using a tail and control mass", Baltimore, MD, June 2010, IEEE, American Controls Conference.
23. C. Orlowski, *Flapping Wing Micro Air Vehicles: An Analysis of the Importance of the Mass of the Wings to Flight Dynamics, Stability, and Control*, PhD thesis, University of Michigan, 2011.
24. J. Delaurier, "An aerodynamic model for flapping-wing flight", *Royal Aeronautical Society*, pp. 125–130, 1993.
25. J. Dietl and E. Garcia, "Stability in ornithopter longitudinal flight dynamics", *Journal of Guidance, Control, and Dynamics*, vol. 31, no. 4, pp. 1157–1162, July–August 2008.
26. J. Lee, J. Kim, D. Kim, and J. Han, "Longitudinal flight dynamics of bio-inspired ornithopter considering fluid-structure interaction", Toronto, Ontario Canada, August 2010, AIAA, Atmospheric Flight Mechanics Conference.
27. D. Kim, S. Lee, and J. Han, "Improved aerodynamic model for efficient analysis of flapping-wing flight", AIAA *Journal*, vol. 49, no. 4, pp. 868–872, April 2011.
28. J. Lee, J. Kim, D. Kim, and J. Han, "Longitudinal flight dynamics of bioinspired ornithopter considering fluid-structure interaction", *Journal of Guidance, Control, and Dynamics*, vol. 34, no. 3, pp. 667–677, May–June 2011.
29. D. Kim, H. Kim, J. Han, and K. Kwon, "Experimental investigation on the aerodynamic characteristics of

a bio-mimetic flapping wing with macro-fiber composites", *Journal of Intelligent Material Systems and Structures*, vol. 19, no. 3, pp. 423–431, 2008.

30. M. Vest and J. Katz, "Unsteady aerodynamic model of flapping wings", AIAA *Journal*, vol. 34, no. 7, pp. 1435–1440, July 1996.
31. B. Roget, J. Sitaraman, R. Harmon, J. Grauer, J. Hubbard, and S. Humbert, "Computational study of flexible wing ornithopter flight", *Journal of Aircraft*, vol. 46, no. 6, pp. 2016–2031, November–December 2009.
32. W. Shyy, Y. Lian, J. Tang, D. Viieru, and H. Liu, *Aerodynamics of Low Reynolds Number Flyers*, Cambridge Aerospace Series. Cambridge University Press, 2007.
33. D. Keats-Pullen, J. Hubbard, and N. Guerreiro, "Spatial weighting of smart materials for real-time measurement of aerodynamic forces", *Journal of Intelligent Material Systems and Structures*, vol. 19, no. 7, pp. 837–844, July 2007.
34. M. Bolender, "Rigid multi-body equations-of-motion for flapping wing mavs using kane's equations", Chicago, IL, August 2009, AIAA, Guidance, Navigation, and Control Conference.
35. K. Sibilski, "Dynamics of micro-air-vehicle with flapping wings", *Acta Polytechnia*, vol. 44, no. 2, pp. 15–21, 2004.
36. J. Grauer and J. Hubbard, "Modeling of ornithopter flight dynamics for state estimation and control", Baltimore, MD, June 2010, IEEE, American Controls Conference.
37. V. Klein and E. Morelli, *Aircraft System Identification: Theory and Practice*, AIAA, AIAA Education Series, 2006.
38. D. Doman, M. Oppenheimer, and D. Sigthorsson, "Wingbeat shape modulation for flapping-wing micro-air-vehicle control during hover", *Journal of Guidance, Control, and Dynamics*, vol. 33, no. 3, pp. 724–739, May–June 2010.

39. W. Dickinson, A. Straw, and M. Dickinson, "Integrative model of drosophila flight", AIAA *Journal*, vol. 46, no. 9, pp. 2150–2164, September 2008.
40. X. Deng, L. Schenato, W. Wu, and S. Sastry, "Flapping-flight for biomimetic robotic insects: Part i - system modeling", IEEE *Transactions on Robotics*, vol. 22, no. 4, pp. 776–788, August 2006.
41. S. Chung and M. Dorothy, "Neurobiologically inspired control of engineered flapping flight", *Journal of Guidance, Control, and Dynamics*, vol. 33, no. 2, pp. 440–453, March–April 2010.
42. X. Deng, L. Schenato, and S. Sastry, "Flapping-flight for biomimetic robotic insects: Part ii - flight control design", IEEE *Transactions on Robotics*, vol. 22, no. 4, pp. 789–803, August 2006.
43. G. Taylor and A. Thomas, "Dynamic flight stability in the desert locust schistocerca gregaria", *Journal of Experimental Biology*, pp. 2803–2829, 2003.
44. G. Taylor, R. Bomphrey, and J. Hoen, "Insect flight dynamics and control", Reno, NV, Jan. 2006, AIAA, number 2006–32 in Aerospace Sciences Meeting and Exhibit.
45. I. Faruque and J. Humbert, "Dipteran insect flight dynamics: Part i - longitudinal motion about hover", *Journal of Theoretical Biology*, pp. 1–15, 2010.
46. I. Faruque, *Control-Oriented Reduced Order Modeling Dipteran Flapping Flight*, PhD thesis, University of Maryland, 2011.
47. J. Kim, J. Lee, and J. Han, "Limit-cycle oscillation suppression of ornithopter longitudinal flight dynamics", Portland, OR, Aug. 2011, AIAA, number 2011–6404 in Guidance, Navigation, and Control Conference.
48. J. Lee, J. Kim, and J. Han, "Cycle-averaged mean and zero-mean time-varying flight dynamics of the ornithopters at trim", Smart Systems and Structures Lab, KAIST, 2010.
49. G. Taylor and R. Zbikowski, "Nonlinear time-periodic models of the longitudinal flight dynamics of desert

locusts shistocerca gregaria", *Journal of the Royal Society Interface*, vol. 5, pp. 197–221, May 2005.

50. M. Bolender, "Open-loop stability of flapping flight in hover", Toronto, Ontario Canada, August 2010, AIAA, Guidance, Navigation, and Control Conference.
51. J. Dietl and E. Garcia, "Ornithopter control with periodic infinite horizon controllers", *Journal of Guidance, Control, and Dynamics*, vol. 34, no. 5, pp. 1412–1422, Sep.–Oct. 2011.
52. R. Tedrake, Z. Jackowski, R. Cory, J. Roberts, and W. Hoburg, "Learning to fly like a bird", *under review*, 2010.
53. S. Kinkade, "Hobby technik", www.hobbytechnik.com, 2010.
54. J. Grauer and J. Hubbard, "Inertial measurements from flight data of a flapping-wing ornithopter", *Journal of Guidance, Control, and Dynamics*, vol. 32, no. 1, pp. 326–331, Jan.–Feb. 2009.
55. T. Gainer and S. Hoffman, "Summary of transformation equations and equations of motion used in free-flight and wind tunnel data reduction and analysis", Tech. Rep. SP–3070, NASA, 1972.
56. E. Morelli, "High accuracy evaluation of the finite fourier transform using sampled data", Tech. Rep. 110340, NASA, June 1997.
57. J. Sitaraman and J. Baeder, "On the field velocity approach and geometric conservation law for unsteady flow simulations", 2003, number 2003–3835 in Applied Aerodynamics Conference.
58. M. Shuster and S. Oh, "Three-axis attitude determination from vector observations", *Journal of Guidance and Control*, vol. 4, no. 1, pp. 70–77, 1981.
59. J. Slotine and W. Li, *Applied Nonlinear Control*, Prentice Hall, 1991.
60. E. Whittaker, *A Treatise on the Analytical Dynamics of Particles and Rigid Bodies*, Cambridge University Press, 1917.

61. D. Greenwood, *Advanced Dynamics*, Cambridge University Press, 2003.
62. F. Lewis, D. Dawson, and C. Abdallah, *Robot Manipulator Control: Theory and Practice*, Marcel Dekker, 2004.
63. R. Kelly, V. Santibáñez, and A. Loría, *Control of Robot Manipulators in Joint Space*, Springer, 2005.
64. R. Ortega and A. Loria, *Passivity-Based Control of Euler-Lagrange Systems*, Springer, 1998.
65. J. Brewer, "Kronecker products and matrix calculus in system theory", IEEE *Transactions on Circuits and Systems*, vol. 25, no. 9, pp. 772–781, September 1978.
66. I. Fantoni and R. Lozano, *Nonlinear Control for Underactuated Mechanical Systems*, Springer, 2002.
67. L. Ljung, *System Identification: Theory for the User*, Information and System Sciences Series. Prentice Hall, 1999.
68. M. Tischler and R. Remple, *Aircraft and Rotorcraft System Identification: Engineering Methods with Flight Test Examples*, AIAA Eduation Series. AIAA, 2006.
69. V. Klein, J. Batterson, and P. Murphy, "Determination of airplane model structure from flight data using modified stepwise regression", Tech. Rep. TP–1916, NASA, October 1981.
70. R. Maine and K. Iliff, "Application of parameter estimation to aircraft stability and control: The output-error approach", Tech. Rep. TM 1168, NASA, 1986.
71. Anonymous, "Flying qualities of piloted aircraft", Tech. Rep. MIL–STD–1797, US Department of Defense, 1995.
72. J. Grauer, J. Conroy, J. Hubbard, J. Humbert, and D. Pines, "System identification of a miniature helicopter", *Journal of Aircraft*, vol. 46, no. 4, pp. 1260–1269, July–Aug. 2009.
73. J. Grauer, E. Ulrich, J. Hubbard, D. Pines, and J. Humbert, "Model structure determination of an ornithopter aerodynamics model from flight data", Orlando, FL, January 2010, AIAA, Aerospace Sciences Meeting.

74. D. Montgomery, E. Peck, and G. Vining, *Introduction to Linear Regression Analysis*, Wiley, 2001.
75. D. Belsley, E. Kuh, and R. Welsh, *Regression Diagnostics: Identifying Influential Data and Sources of Collinearity*, Wiley, 1980.
76. Y. Xiong and M. Sun, "Stabilization control of a bumblebee in hovering and forward flight", *Acta Mech Sin*, vol. 25, pp. 13–21, 2008.
77. H. Khalil, *Nonlinear Systems*, Prentice Hall, 2011.
78. W. Rugh, *Linear System Theory*, Information and system sciences. Prentice Hall, second edition, 1996.
79. P. Dorato, C. Abdallah, and V. Cerone, *Linear Quadratic Control: An Introduction*, Krieger, 2000.
80. D. Gebre-Egziabher, G. Elkaim, D. Powell, and B. Parkinson, "Calibration of strapdown magnetometers in magnetic field domain", *Journal of Aerospace Engineering*, vol. 19, no. 2, pp. 87–102, April 2006.
81. Q. Li, "Least squares ellipsoid specific fitting", 2004, IEEE paper No. 0–7695–2078–2.
82. J. Crassidis, F. Markley, and Y. Cheng, "Survey of nonlinear attitude estimation methods", *Journal of Guidance, Control, and Dynamics*, vol. 30, no. 1, pp. 12–28, January–February 2007.
83. E. Morelli, "Real-time dynamic modeling: Data information requirements and flight test results", *Journal of Aircraft*, vol. 46, no. 6, pp. 1894–1905, Nov.–Dec. 2009.
84. L. Meirovitch, *Fundamentals of Vibrations*, McGraw-Hill, 2001.
85. T. Tarn, A. Bejczy, X. Yun, and Z. Li, "Effect of motor dynamics on nonlinear feedback robot arm control", IEEE *Transactions on Robotics and Automation*, vol. 7, no. 1, pp. 114–122, February 1991.
86. L. Nguyen, M. Ogburn, W. Gilbert, K. Kibler, P. Brown, and P. Deal, "Simulator study of stall/post-stall characteristics of a fighter airplane with relaxed longitudinal static stability", Tech. Rep. TP 1538, NASA, 1979.

Index

CPSIA information can be obtained at www.ICGtesting.com
Printed in the USA
LVOW10*1443021213

363566LV00010B/240/P

9 780857 094667